A SURVEY OF RADIAL VELOCITIES
IN THE
ZODIACAL DUST CLOUD

Brian Harold May
Astrophysics Group
Department of Physics
Imperial College, London

Thesis accepted for the degree of Doctor of Philosophy by the
Imperial College of Science, Technology and Medicine
London 2007

Declaration and Copyright

With reference to my 1970s publications on this work, in *Nature* and MNRAS (see Appendices 5 and 6), please note that, following the tradition in Prof. James Ring's Infrared Astronomy Department, we credited these papers alphabetically.

A C.I.P. Catalogue record for this book is available from the Library of Congress.

ISBN 978-0-387-77705-4 (HB)

Published by Springer 2008
233 Spring St
New York, NY 10013, USA

In association with
Canopus Publishing Limited
27 Queen Square, Bristol BS1 4ND
www.canopusbooks.com

Producer: Robin Rees
Editing and layout: Julian Brigstocke, Tom Jackson
Proofreader: Jennifer Tunney

Printed by Short Run Press, Exeter, UK.

This thesis is dedicated to Harold May.

This was your dream too, Dad.

Figure 0.1 The Zodiacal Light, photographed by the author in Izaña, April 1971.

"Dust in the Wind … all we are is Dust in the Wind …"

(Lyric by Kerry Livgren, of *Kansas*, 1977, from the album "Point of No Return")

Figure 0.2 View from the Observatory site - showing Mount Teide, and the Sea of Clouds. Photo: Brian May, 1971.

Foreword

In the summer and autumn of 2006 I read several interviews with Brian May in which he mentioned his desire to complete the PhD that he had abandoned in 1974. I looked up the papers he had published while a PhD student, which were on spectroscopic studies of the motion of the dust responsible for the zodiacal light, and felt that there was a basis for a thesis. Since he had been a student at Imperial, I knew, as Head of the Astrophysics Group at Imperial, that it would be good for the Group if he came and worked with us. I got in touch with him by email and suggested he come and talk about it. He replied enthusiastically and said that he was working on typing up what he had completed by 1974. I gradually realized that I was the only staff member at Imperial who had previously worked on zodiacal dust, so that I would have to act as his supervisor. Eventually we met and I tried to assess whether he would be able to find time for the huge amount of work that finishing off a thesis involves, particularly if it has not been touched for over 30 years. Since some of Brian's emails were coming from the recording studio I knew there was strong competition for his time.

I sketched out for Brian the work on zodiacal dust in the intervening years that he would have to review and compare with what he had done in the 70s. We did at first plan to start a new programme of observations to go in the thesis, but eventually decided to leave this for the future. He worked really hard at the thesis, though I would sometimes get emails saying, "I can't go on. This is killing me." Not surprisingly he did find it very hard to get back into astrophysics research after such a long gap.

Brian is a great perfectionist. The Imperial astrophysics postgraduates write their theses with a latex macro which passes from generation to generation. Brian insisted on writing in Word and with infinite pain managed to generate a final result that looked just like the standard thesis. We asked him to give a talk on zodiacal dust for a conference in July 2007 for my 65th birthday. Brian had not used Powerpoint before, so I sent him a short tutorial. Brian came up with a brilliant talk which included an animation of the solar system, complete with comets, asteroids and zodiacal dust, and accompanied by a Brian May guitar solo. Of course his techies had to come in first and check the sound system.

It was at this conference that his PA said to me, "Can we have him back now?"

Brian showed incredible determination in completing this thesis, which has moved a long way from the abandoned project of 1974. He has almost certainly set a new record of 37 years between registration and final submission and I doubt that this will ever be broken. From August 1st, 2007, Imperial College broke away from the University of London and started to award its own PhDs. Brian was determined for this to be the first Imperial College PhD thesis and was on the doorstep of the Registry at 9am on August 2nd. The oral was nerve-wracking but went fine and the thesis was accepted with only minor corrections to be made. Brian's fame as a guitarist and mastermind of Queen means that there has been a lot of interest in this thesis. He has single-handedly raised the profile of zodiacal dust in astronomy and has been in contact with many of the leading workers in the field. He plans to continue with research and we have made him a Visiting Postdoctoral Researcher at Imperial. Plans include a new programme of observations at Tenerife and an international conference on dust in the Solar System in 2009.

It has been great fun working with Brian on this project over the past year. He is a very charming and modest man and has immense enthusiasm. He is probably the last PhD student I will take on and certainly the most unusual. Because this is a record-breaking thesis, and is the first ever Imperial PhD, it is appropriate that it should be published. For the non-technically minded it includes a marvellous history of the zodiacal light and its interpretation, as well as a very amusing account of his time on Tenerife in the 1970s making these observations.

Professor Michael Rowan-Robinson, 2007

Figure 0.3 Brian inside the crater at the very top of Mt. Teide in 1971 - some way above the spot chosen by Piazzi Smyth for his second site-testing base at Alta Vista.

Preface and Acknowledgements

This thesis documents a body of spectroscopic observational work on the Zodiacal Light, which was completed 30 years ago. The results were published in two papers (see Appendices) in Nature, 1972, and in Monthly Notices of the Royal Astronomical Society, 1974. The project was devised and adopted under the guidance of Prof. James Ring and Dr. Ken Reay, of the Department of Infrared Astronomy, Imperial College, London. On my first observing trip I was accompanied by Dr. Reay, and on my second, by a young fellow research student, Tom Hicks. For these publications, we made the decision to credit our names alphabetically, as *Hicks, May and Reay*, but it is important to note here that the conduct of this research was in my hands; it was always understood that this body of work was my own project, the substance of my PhD. Alongside this clarification, my respects and thanks are here gladly acknowledged to my colleagues in this venture. I inherited a Fabry-Perot spectrometer and pulse-counting equipment from Prof. Ring, and spent 18 months entirely rebuilding and updating both the optics and electronics, in preparation for obtaining essentially the first viable set of radial velocity measurements, all around the ecliptic, of the Zodiacal Light. The writing of my thesis was virtually complete in 1974, but the submission was deferred due to various pressures. Having returned to the work in 2006, I am happy to submit a complete thesis in 2007, in which I am now able to offer some new perspectives, because of the research which has taken place between 1974 and the present day.

Although new observations are currently being undertaken, the core of this thesis is my observational work conducted between 1971 and 1974. It furthers a line of inquiry initiated in 1962 by James Ring and others at Mt. Chacaltaya in the Bolivian Andes.

The three years (eventually extended to four) of my initial appointment at Imperial College Infrared Astronomy Department involved me firstly in an attempt to make observations from Testa Grigia (altitude 11,500 ft) in the Italian Alps, totally frustrated by atrocious weather, and subsequently in the retrieval of the equipment from under six feet of snow, and its transport back to England. I then concerned myself with the updating of a spectrometer with its control and read-out equipment, the building of a small observatory (hut!) to put it in, on a volcanic ridge in Tenerife, and three seasons of on-site work. The first visit was spent largely in discussing the mysteries of English blueprints with Tenerife builders, but observations during the remaining two did indeed provide the first comprehensive body of high-resolution spectra of the Zodiacal Light ever amassed. Two years of work on the data followed, revealing an unexpected discovery of Magnesium emission, as well as much new information on the solar Zodiacal Cloud.

I was most fortunate, in the initial phase (1970-4), to work under the wing of the following two experienced Zodiacalists.

My grateful thanks to my initial supervisor, the late Professor James Ring, in the first place for taking me on as a research student in the Infrared Astronomy Department of Imperial College, London, subsequently for his constant interest, faith, and advice, and later for the very great favour, to someone no longer strictly a College member, of continued access to the life and amenities of the Imperial College Infrared Astronomy Group in the 1970s. Also to my co-supervisor Dr. Ken Reay, who gave unsparingly of his time and energy during the first four years of my studies, at 10, Prince's Gardens, South Kensington.

My long neglected gratitude to my Dad, Harold May, who strongly supported my scientific endeavours all through my life, until he passed away too soon, and too soon to see the full impact of his guidance and example to me. He also helped with some of the diagrams and figures, and his influence, his immaculate attention to symbolic detail and stylish draughtsmanship, can still be seen in some of the graphics in this thesis. I hope I will at last do him justice. And to my dear Mum, Ruth, who encouraged me in all aspects of my life, and gave me perhaps the deepest parts of my humanity.

I am deeply indebted to Prof. Francisco Sanchez Martinez of the University of La Laguna, Tenerife, for the use of the facilities of the Observatorio del Teide in the 1970s, and for his encouragement and support during my visits to Tenerife as a young student; to my dear friend, now sadly departed, Dr. Carlos Sanchez Magro, who indispensably organised local materials, equipment and labour, and taught me to swear in Spanish! I hope you will be proud, finally, Carlos, of this fruit of our friendship.

To my successor in spectral radial velocity work, Tom Hicks of Imperial College, who shared part of the observational work, helped in the data reduction, and from whom I have benefited in many discussions; to J. E. Crabtree, W. Stannard and A. V. Demenezes for help and indulgence in constructional matters; to Isaac Newton (the departmental cat, now sadly departed), my sole comfort and inspiration during the seemingly interminable initial reduction of an enormous pile of data; to the Science Research Council for financially supporting the first three years research, and to Stockwell Manor School (where I taught Maths as a supply teacher to fund a further year of research) and "Queen", for support and distraction during the final year; to Miss Christine Mullen (later Mrs. May) for much help and encouragement to sustain the work in circumstances which were sometimes difficult, and to Miss Mary Davies (now Mrs. Mary Lewis), who typed the first thesis manuscript, now lost.

Returning to the study of the Zodiacal Light in 2006, I have based this new version directly on my 1970s handwritten notes, and in most places it adheres closely to the original text, updated only where qualification was necessary in the light of subsequent experience. In this second phase, I have benefited from new perspectives and expert advice of the highest quality, and I offer my sincere thanks:

To Prof. Michael Rowan-Robinson, Head of Astrophysics, Imperial College, London, who has done me the great honour, as my new supervisor, of enabling me to re-register for a PhD at Imperial College, offering me the opportunity finally to complete a Great Circle in my life. Also for giving me, after an absence of 30 years, a 'home' in the Blackett building; and above all, for his caring and careful criticism, tutoring and advice, which have lifted this thesis to Gold Standard.

Once again to Prof. Francisco Sanchez Martinez, a hero to me for 35 years, who has re-opened the doors of the Instituto de Astrofisica de Canarias (IAC), to enable me once again to take advantage of the facilities in the Observatorio del Teide - the place where this line of research all began - for Charles Piazzi Smyth in 1853, and for me in 1971.

To Dr. Garik Israelian, my prime collaborator in resuming this work, who has paid me the huge compliment of putting his own research programmes on hold, in order to plunge into a new Zodiacal Cloud study, and is more than anyone else responsible for helping me through the final stages of this PhD work. Also to his lovely family.

To Sir Patrick Moore, who inspired me a lifetime ago to be an astronomer in the first place, for his personal encouragement to rebuild my participation in Astronomy, particularly during a three-year period whilst we collaborated on a popular book on the History of the Universe - *Bang!* - motivating my return to Zodiacal Light research.

To Dr. Chris Lintott, our third collaborator in the Bang! book, for educating me in the ways of modern astronomical communication via the Internet, for a crash course in a whole spectrum of new cosmological concepts, and for many suggestions.

My thanks also, for encouragement and advice in so many areas, to Tom Short, of the University of Herts., my best friend ever since my old college days, who has been kind enough to check my manuscript; to Prof. Stan Dermott for many invaluable insights; to Dr. Tom Kehoe, Ashley Espey, Prof. Ian Robson, Marcus Chown, Robin Rees, Brian Smallwood, Prof. Tom Gehrels, Yvonne Powell, Carey Lisse, Dr. Noah Petro, Pete Malandrone, Richard Gray, and Prof. Greg Parker; to Jennifer Tunney for truly indispensable proof-reading; to Sara Bricusse and Claire Bartlett for timetabling and personal backup, and Philip Webb for low-stress transportation; to Justin Shirley-Smith for invaluable computing support; to Prof. Kirpal Nandra, Prof. Rees Rawlings, Prof. Janet Drew, Juliet Kemp, Loli Sanchez, Henry Jestico, Dr. Elise Laird, Dr. David Clements, and Diana Moore, all of Imperial College, for smoothing my way back into college life; and to the examiners of this thesis, Dr. Iwan Williams and Dr. Mark Jones, who, by their careful analysis, significantly improved my text. To my dear friends Elena Torrents and Jose Manuel Vergara, and Inge and Patricio Feier, for giving me a spiritual home in Tenerife; to my children, Jimmy, Louisa and Emily, for their great enthusiasm for the project, in spite of its all-consuming demands on their Dad's already stretched availability; and to my dear wife, Anita, the essential pivot and inspiration of my life, without whom I would not have had the will to journey into new lands.

Brian May, 2007

In the reference research for this thesis I have made use of NASA's Astrophysics Data System Bibliographic Services (ADS). Thanks! I would also like to acknowledge those two new, though somewhat lawless, portals to the World's knowledge: Google, and Wikipedia.

Figure 0.4 Growth of the observatory in Tenerife over 35 years. *Above:* The Observatorio del Teide in Sept 2006. *Below:* View of the ZL Hut (centre) and the site at Izaña, looking east from the same viewpoint, 1971.

Contents

Abstract

This thesis documents the building of a pressure-scanned Fabry-Perot Spectrometer, equipped with a photomultiplier and pulse-counting electronics, and its deployment at the Observatorio del Teide at Izaña in Tenerife, at an altitude of 7,700 feet (2,567 m), for the purpose of recording high-resolution spectra of the Zodiacal Light. The aim was to achieve the first systematic mapping of the MgI absorption line in the Night Sky, as a function of position in heliocentric coordinates, covering especially the plane of the ecliptic, for a wide variety of elongations from the Sun. More than 250 scans of both morning and evening Zodiacal Light were obtained, in two observing periods – September-October 1971, and April 1972. The scans, as expected, showed profiles modified by components variously Doppler-shifted with respect to the unshifted shape seen in daylight. Unexpectedly, MgI emission was also discovered. These observations covered for the first time a span of elongations from 25° East, through 180° (the Gegenschein), to 27° West, and recorded average shifts of up to six tenths of an angstrom, corresponding to a maximum radial velocity relative to the Earth of about 40 km/s. The set of spectra obtained is in this thesis compared with predictions made from a number of different models of a dust cloud, assuming various distributions of dust density as a function of position and particle size, and differing assumptions about their speed and direction. The observations fit predominantly prograde models fairly well, but show a morning-evening asymmetry, different in the two observing periods. Models are investigated containing various components, including prograde and retrograde orbiting dust around the Sun, a drift of interstellar material though the Solar System, and light from distant emitting matter. The implications for possible asymmetries of the Zodiacal Cloud are discussed. Other researches on the Zodiacal Dust Cloud, before, during, and after my observations, are reviewed, including recent insights into its structure, orientation, and evolution, up to the present day, and my observations are evaluated in this context. Period of study, 1970–2007.

Figure 0.5 The end of the night. The coelostat silhouetted against the Sunrise.

List of Figures

Figure 0.6 View from the Observatory to the East, just after Sunset, showing the shadow of Mt. Teide, and a full Moon.

Abbreviations and Acronyms

AU ...... Astronomical Unit
CIB ...... Cosmic Infrared Background
CIBER ...... Cosmic Infrared Background ExpeRiment
CTC ...... Cis-Terrestrial Cloud
CMB ...... Cosmic Microwave Background
COBE ...... Cosmic Background Explorer
Dec ...... Declination
DIRB ...... Diffuse Infrared Background
DIRBE ...... Diffuse Infrared Background Experiment
EHT ...... Extra High Tension
F-P ...... Fabry-Perot (interferometer)
FIRAS ...... Far Infrared Absolute Spectrophotometer
GDC ...... Geocentric Dust Cloud
HMR ...... Hicks, May and Reay
IDP ...... Interplanetary Dust Particle
IPD ...... Interplanetary Dust
ISM ...... (Galactic) InterStellar Medium
IRAS ...... Infrared Astronomical Satellite
ISO ...... Infrared Space Observatory
KBO ...... Kuiper Belt Object
LIC ...... Local Interstellar Cloud
MDTL ...... Motorola Diode Transistor Logic
MNRAS ...... Monthly Notices of the Royal Astronomical Society
NPL ...... National Physical Laboratory
PTFE ...... Polytetrafluoroethylene
RA ...... Right Ascension
RHVW ...... Rowan-Robinson *et al* (1990) model of the Zodiacal Cloud
WHAM ...... Wisconsin Hydrogen Alpha Mapping Spectrometer
ZDC ...... Zodiacal Dust Cloud
ZC ...... Zodiacal Cloud
ZL ...... Zodiacal Light

Chapter 1
Introduction

The scope of this introductory chapter is to review the entire history of Zodiacal Light research, from the earliest times, through the period of my own original research, right up to the present day. Particular attention is paid to the development of Doppler measurements, and to the most recent revelations of infrared surveys.

1.1 The Zodiacal Light – the phenomenon

The Zodiacal Light, conveniently abbreviated to 'ZL', is the name given to a diffuse cone of light seen in the Western sky in the evening, from around the end of twilight, and in the Eastern morning sky just before dawn. In the latter incarnation it has been called the 'False Dawn'. If it could be viewed in a completely dark sky when the Sun was eclipsed, we might imagine that the complete outline of a brightness contour of the ZL would be an ellipse, with the Sun at its centre, and its long axis roughly aligned with the ecliptic. To see this phenomenon clearly on Earth, it is necessary to have a good, dark sky, free from clouds, terrestrial light sources, moonlight, or aurorae. It is also an advantage to be within or near the tropics, since it is here that the cone of light is inclined at a high angle to the horizon, making it still visible when the Sun is well below the horizon, and the sky is completely dark. The maximum intensity of the ZL is greater than the brightest parts of the Milky Way, and in ideal conditions, the morning and evening cones are seen to be linked up, by a continuous faint band right around the ecliptic. This 'Zodiacal Band' is visible all through the night, and shows a secondary brightening at the anti-solar point, known as the "Gegenschein". With modern instruments more sensitive than the human eye, it has become apparent that even well off the ecliptic, indeed even at the ecliptic pole, there is still some contribution to the sky background brightness from this phenomenon. The Zodiacal Light, in fact, surrounds us.

A knowledge of the nature and distribution of the material responsible for the ZL is interesting for many reasons. At the time when I began this research, around 1970, manned space flight was just beginning, and there was a need to evaluate what dangers to astronauts there might be from high-energy micrometeorites. In studies of the origin of our Solar System, too, there are benefits in knowing the composition and motions of the interplanetary dust; the present parameters of this cloud of material can give indications as to how our planetary system was formed, and how the Sun's environment is evolving at present. In the 21st Century, studies of the ZL have become relevant for new reasons; dozens of exo-planets in other solar systems are now known to us in 2007, but as early as 1984 in the IRAS experiment, the "Vega Phenomenon" was discovered (see discussion by Beichman 1987) – infrared excesses in radiation indicating clouds of solid grains around three nearby Main Sequence stars, Alpha Lyrae (Vega) (Aumann *et al* 1984), Alpha Piscis Austrinus (Fomalhaut), and Beta Pictoris. Many other discoveries followed, and it now appears that these exo-circum-stellar dust clouds – the Zodiacal aggregations of other suns - are very common features of the environs of stars. The most studied of the exo-zodiacal clouds is perhaps *ß* Pictoris (Smith, Terile 1984), whose dust debris disc, viewed conveniently edge-on from Earth, appears to be concentrated in more than one plane, perhaps indicating disturbance from yet undetected planets (Hubble website 2006). A thorough knowledge of our own evolving system of planets, debris and dust helps to shed light on the evolution of planetary systems in general. Of course the search is highlighted by a desire to answer the question of whether our Earth, and life as we know it, is unique in Creation, or if such life is scattered liberally throughout the Universe. Zodiacal Dust may still be holding on to some of the secrets relating to these questions.

1.2 Review of observations of the ZL prior to 1970

1.2.1 Earliest views

"If I take the wings of the morning, and dwell in the uttermost parts of the sea, even there shall thy hand lead me, and thy right hand shall hold me" - Psalms 39, vv. 9-10.

This biblical quotation has been mooted (Fechtig *et al* 2004) as a possible reference to the appearance of the morning ZL, but other translations give the verse a different meaning: the *'Word of God'* translation has it as *"If I climb upward on the rays of the morning Sun, or land on the most distant shore of the sea where the Sun sets ..."* Moreover it is common for wings to be poetically ascribed to the Sun and Moon, and also to the Morning itself, which may be the case here. So there is some doubt as to whether these words from the Bible's Book of Psalms actually *do* indicate an awareness of the Zodiacal Light in Old Testament times. Are there any other signs of ancient knowledge of the ZL? Fechtig *et al* (2004) allow that the Aztecs (16th Century) may be correctly credited with having noticed the ZL, and that there are possible references from the Chaldeans (600 B.C.) and the ancient Egyptians, but such occasional mentions are few indeed compared with the liberal scattering of allusions in art of all kinds to other heavenly manifestations (for example, the works of Shakespeare contain a virtual catalogue of

DÉCOUVERTE
DE LA LUMIERE
CELESTE
QUI PAROIST
DANS LE ZODIAQUE.
PAR MONSIEUR CASSINI
de l'Academie Royale des Sciences.

A PARIS,
DE L'IMPRIMERIE ROYALE,
Par SEBASTIEN MABRE-CRAMOISY, Imprimeur de Sa Majeſté
& Directeur de ſon Imprimerie Royale.

M. DC. LXXXV.

Figure 1.1 Cassini's paper of 1685, claiming the discovery of the Zodiacal Light.

atmospheric phenomena!). The great exception seems to be Omar Khayyam, the great astronomer-poet of Persia. In his Rubaiyat (*circa* 1120), there are two passages highly evocative of the ZL:

"Awake! For Morning in the Bowl of Night
Has flung the Stone that puts the Stars to Flight" and ...

"When False Dawn streaks the East with cold, gray line
Pour in your cups the pure Blood of the Vine
The Truth, they say, tastes bitter in the mouth,
This is a token that the "Truth" is wine."

In the West, the scientific 'discovery' of the ZL is attributed to Giovanni Domenico Cassini, the famous French astronomer, who, in his publications of 1683 and 1685 (see Figure 1.1) claims the find.

Certainly he was the first to quantify it, chart it in the celestial sphere, and, with great insight, suggest an explanation for the phenomenon. But 22 years earlier, J. Childrey (1661) had already brought the ZL to the attention of the Western world. In his work, "Britannica Baconica, or the Natural Rarities of England, Scotland, and Whales (sic)" he wrote, *"In February you shall see a plainly discernible way of the twilight striking up towards to the Pleiades ... and I believe it ... will be constantly visible at that time of the year. But what the cause of it in Nature should be I cannot yet imagine but leave it to further enquiry"*. This is the classic view of the evening Zodiacal Light, and to this day, February - March remains the perfect time of year to see it in the Northern Hemisphere. Cassini and his student Niccolo Fatio de Duillers went a long way towards 'explaining' the appearance of the Zodiacal Light – even suggesting dust particles in orbit around the Sun as the source. But the fact that these men were the first to document the extent of this quite spectacular manifestation in the Heavens is in itself curious.

It is remarkable that so few earlier records are available, since, unless a sudden dramatic increase in brightness occurred in the seventeenth century, the mysterious Light of the Zodiac ought to have been very conspicuous in the dark skies of earlier civilisations, comparable in dimensions with the Milky Way, and at its brightest, more luminous. The number of people living in the modern world who will ever see the Zodiacal Light must be dwindling fast, as urbanisation spreads with its accompanying light pollution, but in ancient times the unpolluted night skies familiar to shepherds and sailors ought to have yielded at least a frequent mention of this ghostly cone of light; and why not a whole folklore, such as pertains to the Milky Way, especially since much of the ancient civilised world resided in or near the tropics? The very fact that there is *no* popular layman's name in any language for what scientists now call the Zodiacal Light suggests that there is an anomaly here. Could it be that our view of the Zodiacal Light is highly variable? Cassini was convinced that it disappeared completely between 1665 and 1681, and this man was certainly no casual observer. I have no solution to the puzzle. In the light of Cassini's reports, along with Jones's observations detailed below, and the curious dearth of references pre-Cassini, I am convinced that, despite the lack of any recent confirmation, we must admit the possibility that the Zodiacal Light has not always been what it is today. One other early set of observations of the ZL is worthy of special mention.

On April 2nd 1853, the Rev. George Jones (1856), Chaplain of the United States Navy, set off on a nautical voyage to Japan and back, and made visual observations of the Zodiacal Light every night, evening and morning, for the whole duration of the trip, a period of over two years. Jones was meticulous in his documentation of the appearance of the phenomenon, and made efforts to eliminate subjectivity from his findings by seeking corroboration from his companions on the ship, who had no prior expectations of what they were supposed to see. His records, published in 1856 in Washington, were the first serious sustained study of the subject, and still represent the biggest single archive of attentions to the Zodiacal Light; in particular, his accounts of its variability with time and latitude still give pause for thought. A page of his charts is reproduced in Figure 1.2, showing Jones's innovative graphic representation of

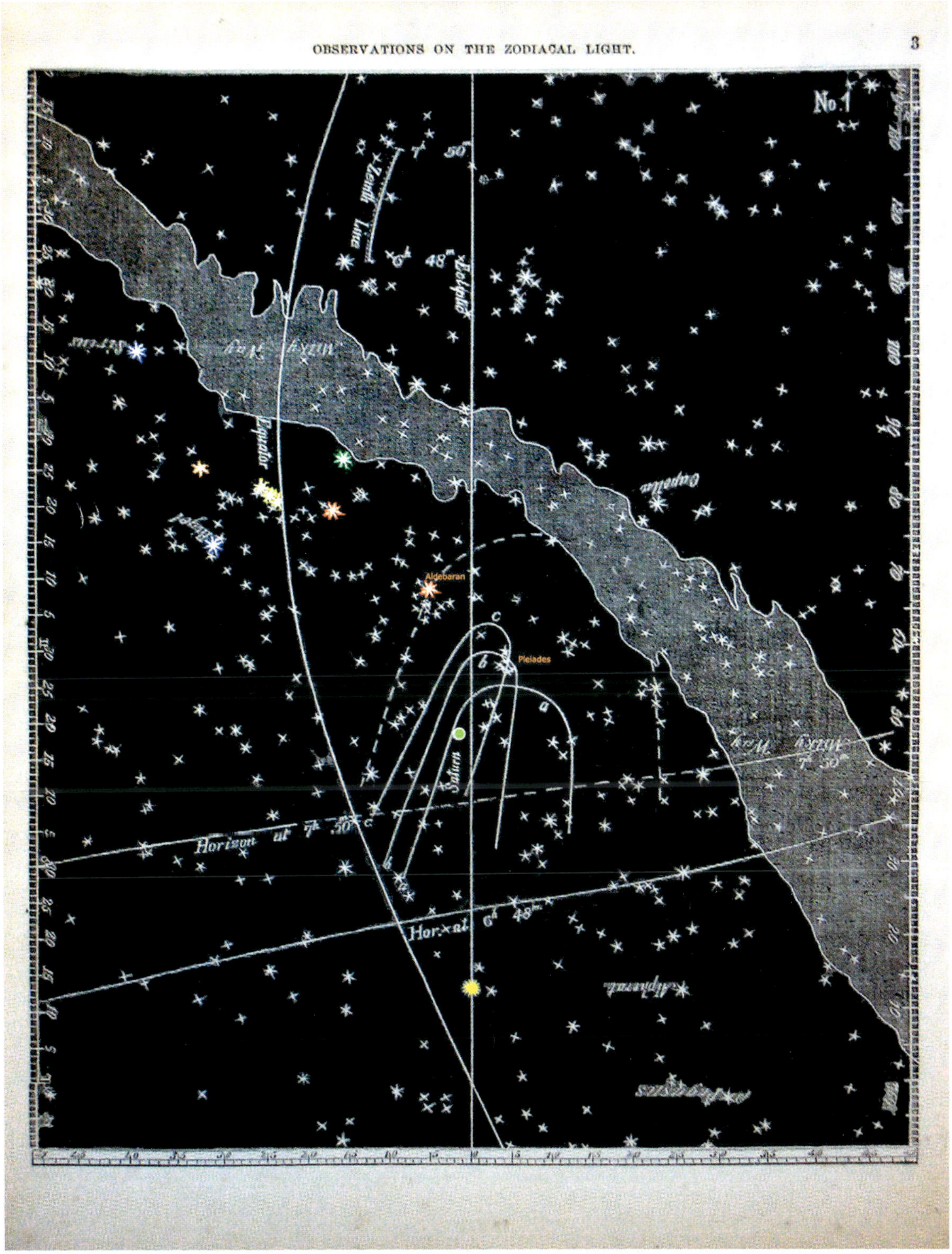

Figure 1.2 A chart from "Observations on the Zodiacal Light from April 2, 1853, to April 22, 1855, made chiefly on board the United States Steam-Frigate Mississippi, during her late cruise in Eastern Seas, and her voyage homeward: with conclusions from the data thus obtained by George Jones, A.K., Chaplain, United States Navy, United States Japan Expedition, Volume III, Washington 1856."

the sky, his own 'projection', built around the ecliptic (vertical in the charts), rather than the celestial equator.

The discovery of the Gegenschein, or counter-glow, is generally attributed to Brorsen, (1854) who first observed it in this year, but he himself acknowledged that Humboldt had named it in 1803, and it is documented that Esprit Pénéza reported its existence to the Paris Academy in 1731. It is interesting to note that Jones did not know of the existence of the Gegenschein until he returned home, but had rejected the observation as unreliable – *because he thought it should not be there* - and only then did he realise that he *had* seen it one night. This is a reminder that, even in such a careful and unbiased observer, the tendency is to see what one expects to see, and *not* see what is not expected. I will have cause to revisit this idea later. Perhaps we can call it the "Lucky Find Syndrome" (LFS, for short, of course!). By 1854, early photography was well under way, but, although scientists of the day (including Piazzi Smyth) had successfully secured images of many astronomical subjects, the sensitised plates of the day, Daguerreotype, Albumen, and Wet Plate Collodion, did not yet possess the sensitivity needed to record an image of the ZL. Hence we only have artistic impressions to go on from this period. A fine engraving of its appearance in 1856, published by Müller in 1865 (see Fechtig *et al* 2004), is reproduced here (Figure 1.3a). The Hyades are shown at top left of the picture, and it will be seen that there are two interlopers to the star scene … two planets, right on the line of the ecliptic (I have coloured them orange), reminding us that the axis of the ZL does indeed lie (roughly) in the ecliptic plane.

Here the Zodiacal cone 'strikes up' towards the Pleiades and Hyades, visible at top left, of the picture, just as in Childrey's description. For comparison, Figure 1.3b (overleaf) shows a photographic view, taken about 150 years later. Although the constellation of Taurus is slightly lower in this view, again the Pleiades and Hyades feature, in a strikingly similar configuration of the Spring ZL, viewed from the Northern Hemisphere. Again, a bright planet is featured.

In all ground-based photographs of the ZL that I have seen, atmospheric extinction modifies the actual recorded shape from a cone to something closer to an ellipse, but this never appears in drawings; it seems that the naked-eye observer instinctively compensates for the effect of darkening at the horizon (and reddening, as seen here).

1856 was also the year in which the celebrated Scottish astronomer Charles Piazzi Smyth (1856) published an account of his expedition to Tenerife, to perform site-testing at high altitudes, the intention being to confirm the theory, first proposed by Newton, that the quality of 'seeing' would improve at

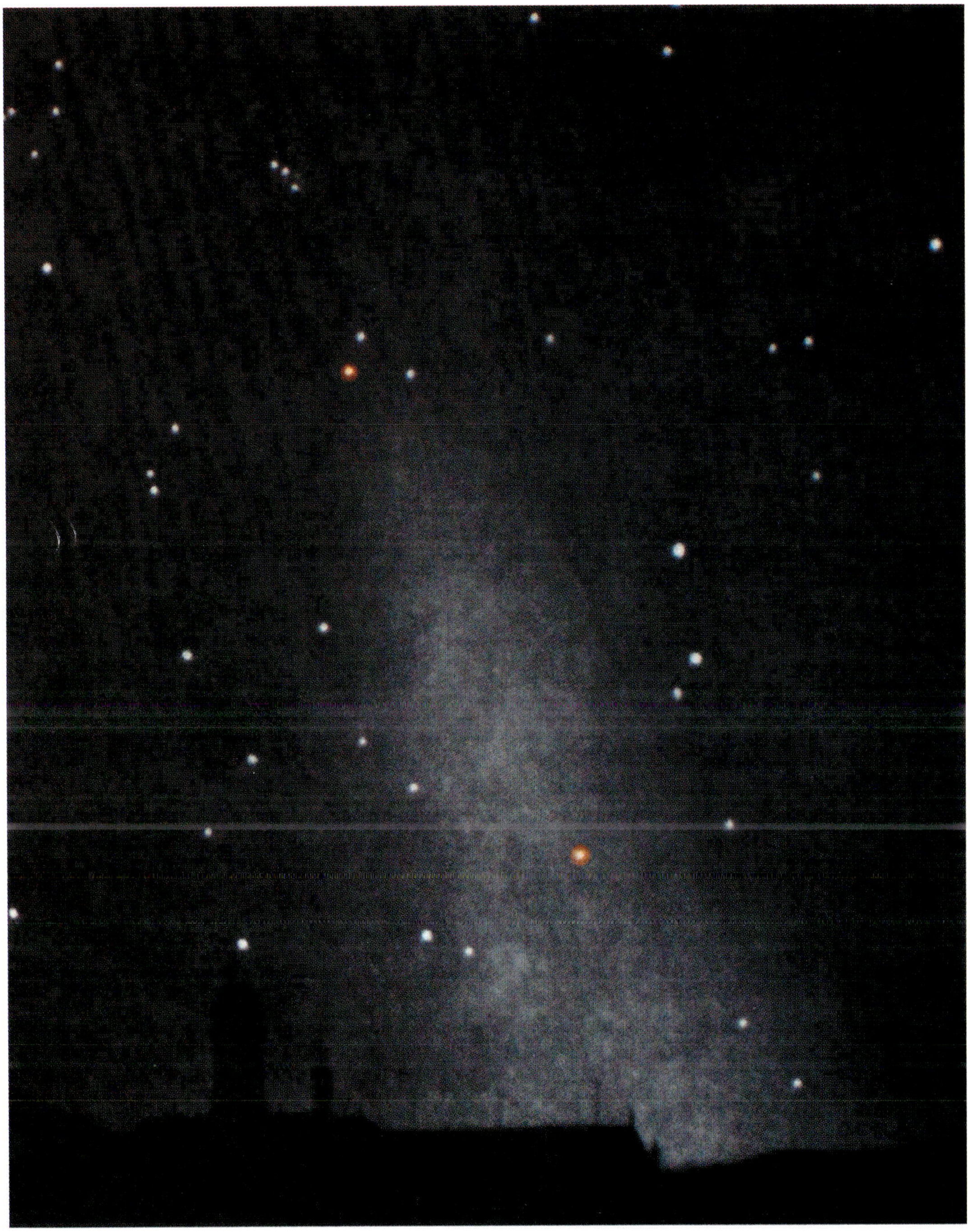

Figure 1.3a The Zodiacal Light as seen on March 5, 1856, over the city of Friebourg in Breisgau, stretching up to the Pleiades, as first described by Cassini; from Müller 1865.

Figure 1.3b The evening Zodiacal Light, photographed by Szymanek in 2006. The Pleiades and Hyades are visible, as described by Cassini in 1865, and illustrated by Müller in1865.

high altitudes. Smyth, one of the first scientists to apply himself to studies of the ZL, suggested that its spectrum must be essentially reflected sunlight, and in Tenerife in 1855 he made the first attempts to confirm this spectroscopically. Over a hundred years later, spectroscopy would be used successfully to map radial velocity patterns relative to the Earth in the Zodiacal Dust Cloud, and I'm sure Piazzi would have been happy to know that this too was to take place in his beloved Tenerife.

1.2.2 Evidence as to the scattering medium

Anders Jonas Ångström, one of the fathers of spectroscopy, discovered the fundamental relationship between absorption lines and emission lines, was the first to find evidence of Hydrogen in the spectrum of the Sun, and gave his name to the Angstrom Unit (10^{-10} m). In 1867 he recorded (incorrectly, it turns out) that he had found the same blue-green emission line in the spectrum of the Zodiacal Light as he had previously (correctly) found in the Aurora Borealis. Since then, it has been substantially confirmed that there are no emission lines present in the visible portion of the spectrum of the ZL. Emission lines from airglow are always liable to be mixed in with ZL measurements, in any ground-based observation, but are now recognisable as such. Medium resolution spectra of the 'cones' (Hoffmeister 1940) showed a continuum with Fraunhofer absorption lines, and Gindilis (1963) found an energy distribution in the Gegenschein the same as sunlight, so it may be regarded as established that the ZL shines by sunlight reflected (or more correctly, scattered) by a relatively slow-moving medium. If electrons were present in the Zodiacal Cloud, due to their high speeds, they would add a highly 'Doppler-smeared' component to the scattered spectrum, filling in the absorption lines. This effect was shown to be absent by Blackwell and Ingham (1961), who, by measuring the depth of Fraunhofer lines, placed an upper limit of $120/cm^3$ on electron density at 1 AU, so the ZL material must be almost exclusively dust. In the light of later ZC models, perhaps this conclusion is now in doubt, since models with orbits that are elliptical, rather than circular, also show shallower absorption lines, and some observers have now claimed to have confirmed this observationally (Madsen *et al* 2006 – see Chapter 4). The composition of the dust was variously estimated to be 'irons' and 'stones' (predominantly silicates), similar to meteorites found on Earth, or graphites, (as some exo-planets are now believed to be), or a mixture of ices (solid carbon dioxide, ammonia, water, etc.), though these would have short lifetimes within a few Astronomical Units from the Sun. The shapes of particles actually collected were, and are, predominantly irregular, or 'fluffy', though a few approximate better to the smooth spheres almost invariably used (Hemenway 1967) in early theoretical models!

1.2.3 Measurable quantities in ZL observations

Measurements of the Zodiacal Light available to various observers up to the 1970's may be summarised according to type as follows:

a) Photometry - Measurements are made of the intensity of the ZL as a function of angular distance ('elongation') from the Sun, and distance from the axis, across the 'cone'. This information can provide pointers to the physical parameters of the cloud, discriminating between theoretical models of particle size and distribution (see, for example, Smith *et al* 1965). Measurements of colour, again as a function of elongation from the Sun, have provided clues to particle sizes (e.g. Elvey 1934). Notably, Carlos Sanchez Magro of the University of La Laguna in Tenerife (PhD Thesis 1974), in a seven-year study, found a dependence of Zodiacal Light brightness on solar activity.

b) Polarisation - Measurements are made of the polarisation of the ZL, usually as a function of elongation from the Sun, an indicator of particle sizes and properties (e.g. Sanchez 1969).

c) Solar Corona measurements - The outer regions of the Solar Corona, so stunningly visible at times of total solar eclipse, merge into the inner parts of the ZL, so measurements on the Corona can be relevant (Beavers *et al* 1980).

d) Position of the axis of the ZL - measurements are made of the angle the axis makes with the ecliptic and its points of intersection. The aim is to ascertain which plane the dust principally lies in, and if its distribution is symmetrical about this plane, and symmetrical East to West. A dependence of the axis position on the latitude of the observer would indicate a GDC component (Geocentric Dust Cloud) – in other words, that a significant proportion of the dust was located near the Earth, rather than in orbits around the Sun. Thus these measurements can enable us to inform an opinion about where the dust lies (e.g. Huruhata, Masaaki 1965). Wolstencroft *et al* (1967) concluded that no GDC component had been detected.

e) Absorption line strength measurement - the spectrum of the Zodiacal Light (sunlight reflected from moving particles) is compared with that of the Sun. The lessening in spectral line strength was believed to indicate a proportion of particles with high velocities 'smearing ' the scattered spectrum (Blackwell and Ingham 1961).

f) Space-borne detection and collection of particles – samples are collected from space vehicles, from the Moon, from deep sea sediments, from permafrost in Antarctica, and from the Stratosphere (see section 1.4.3).

g) Emission (black body) spectra - the warm Zodiacal dust emits radiation in the infrared. This can be detected by ground-based observations, but only through the available atmospheric 'windows'. But this line of enquiry is much easier from a vantage point in space, where there is no restriction on the wavelengths which can be studied. This kind of observation came of age much later, with the IRAS satellite and others (see 1.4.2).

h) Radial velocity measurements - the speed and direction of the motions of dust are inferred from Doppler shifts in the absorption lines in the scattered solar spectrum. This is the line of inquiry pursued in this research project, again principally as a function of elongation from the Sun. Ground-based spectroscopy of the Zodiacal Light, such as that described in this thesis, is well-known to be difficult: there are appreciable challenges to meet – among them, extremely low light levels, atmospheric extinction, airglow, and light pollution from solar and lunar twilight. It would clearly be very attractive to mount a corresponding experiment in space, where these problems would not occur; however, on the ground, experiments can be conducted over a long time period, and much data accumulated.

i) Shapes of Spectral Lines - Clarke *et al* (1995) computed actual profiles due to integrations of Doppler shifts, and were able to predict a bifurcation of minima at elongations close to the Sun. As it becomes possible to observe at higher spectral resolutions, much more could be done in this area.

1.2.4 Physical parameters and models of the ZL

The variables of interest in the interplanetary medium are the number density, size, shape and albedo of the particles, their distance from the Sun, and their speed and direction of motion, almost all of which may be a function of any of the others. In addition, we would like to know the chemical composition of the material, what proportion is gas and what dust, what are its rates of evaporation, melting points, and degree of ionisation and so on. The reason for the slow progress in early ZL research is immediately apparent, since any model constructed to fit a comparatively small range of possible experimental measurements is required to yield values of an enormous number of interconnected physical variables. Simple models of the Zodiacal Cloud are based on number density as a function of position in the Solar System and particle radius.

Some early observers attempted to attribute the existence of the ZL to scattering of light in the Earth's atmosphere, or (as favoured by Jones in 1865) from a dust cloud centred on the Earth; but even in 1683 Cassini had suggested that its appearance might be due to sunlight scattered by a lens-shaped (lenticular) circumsolar cloud of dust concentrated in the plane of the planets, the Earth being situated some way out towards the edge of the 'lens'. This, aside from some exciting recent discoveries about the fine structure of the cloud – zones, or bands, of concentration of material attributed to asteroidal erosion and planetary gravitation - is roughly the view of most astronomers today, though there is still a lingering doubt as to whether a component might be attributed to a Geocentric Dust Cloud (GDC). The Geocentric Theory, though not popular by 1970, still had its adherents (Peale 1966; Divari 1965 and 1966). Early satellite measurements of particle densities seemed to give support to the existence of a significant dust cloud extending to a few Earth radii, giving cause for alarm to those contemplating manned exploration of Space. However, these were later shown to be suspect by Nilsson and Southworth (1967). These researchers, in 1972, detected only TWO impacts in their detector tubes aboard the OGO 2 satellite in hundreds of hours, and the reality of even these was in doubt! So they were able to put an upper limit on micrometeoroids in the vicinity of the Earth of mass greater than 10^{-12} g, of 0.01 particles per cubic metre per second (1967). Indeed these early reports of high concentrations of dust in the region of the Earth were never since confirmed, though we shall see later (1.2.4) that there now seems to be evidence for at least a circumsolar band of dust in the region of the Earth's orbit. Searches for parallax effects in the ZL axis dependent on the observer's latitude yielded conflicting results (perhaps hardly surprising given the difficulty of correcting for atmospheric extinction, and the extreme difficulty of seeing the ZL at all in high latitudes). Roosen (1970) failed to detect the Earth's shadow passing through the Gegenschein, an effect which ought to have been detectable if the dust were near to the Earth (except on one occasion, when a comet was held to be responsible). Roosen also established that the predicted L5 cloud, at the Earth-Moon Libration point, was non-existent. The exhaustive theoretical treatment of Shapiro, Lautman and Columbo (1967) concluded that the existence of a significant GDC is 'unlikely'.

Most researchers now favour circum-solar models, and these have frequently used ZL photometry and polarimetry measurements as their starting points (Behr and Seidentopf 1953; Divari and Asaad 1959; Ingham 1961). Ingham included solar corona data, and fitted parameters to a compromise curve, since there was some difficulty in bridging the two sets of data. The general method, in fitting models to experimental findings, is to make certain simplifying assumptions about the particles (typically spherical particles in circular orbits, with uniform specific gravity and albedo), to restrict attention to the plane of the ecliptic, and to assume values of some parameters, in order to make a suitable fit to the observations by adjustment of the others.

A representative model is one in which the albedo (ratio of reflected light to incident light) of the particles:

$$\gamma = 0.2 - 0.5$$

and the dependence of number density (n) on particle radius (a) and distance from the Sun (r) has the form:

$$n(a,r) = n_0 a^{-p} r^{-\alpha} \qquad \ldots\ldots \quad \text{equation (1.1)}$$

where $p = 3, 4, 5$, and $\alpha = 1, 3/2, 2$.

This is already, evidently, a simplified model, since in reality there is no reason to suppose that p and α are constant, or even orthogonal with each other. The value of n_o is chosen so as to give a total number density of 240 particles per cubic km at 1 AU from the Sun, and a lower limit of 0.3 microns is imposed on the particle radius. This is because very small particles are ejected from the Solar System by the pressure of sunlight (Giese 1961); the value of $a_{\min}$ can in fact be 0.2 microns to 0.4, depending on the density of the particle, which is usually estimated as between 1 and 3 g/cm^3. Further, particles less than 0.2 microns in radius would scatter light selectively according to wavelength, giving rise to predictions of colourations in the ZL which have not been observed, though, again, observations are not all in agreement. In this analysis, the particles are large compared with the wavelength of light under discussion, so the complex theory of Mie Scattering is not needed. Values of p between 0 and 5 were favoured by Weinberg (1965) and also by Ingham (1961), from photometry and polarimetry measurements (see Figure 1.4).

Turning attention to the dependence of n on distance from the Sun, Over (1958) predicted a dust-free zone within about 4 solar radii ($4r_s$) owing to vaporisation by the Sun. Infrared emission peaks were reported by Petersen (1967) supporting the prediction. This can be incorporated in the model by specifying two domains for relation 1.1, an inner one ($r < 4$ solar radii) in which $n(a,r)$ varies as $(4r_s - r)^{-\alpha}$ where α is large, and the outer region where the model is as before (Figure 1.4).

Most *in situ* ZL measurements in this period were made from satellites in the neighbourhood of the Earth, and served to verify and extend the Earth-bound data. Hanner and Weinberg (1973) conducted an interesting experiment in which measurements were made of the Gegenschein intensity as a function of distance from the Sun from an outgoing space vehicle. Their results lent support to the model:

$$n(r) \propto 1/r \quad \ldots\ldots \quad \text{i.e. equation 1.1 with } \alpha = 1.$$

The reason why the Gegenschein is there *at all* was certainly not clear prior to 1970, and may still be in question. If it is not due to a GDC, or to Lunar ejecta streaming outwards in the solar wind (Jeffries 1971), perhaps $n(r)$ in the circumsolar cloud increases outside the Earth's orbit, or perhaps the effect is caused by a back-scattering peak in intensity which might be expected from 'fluffy' particles, and has been predicted from models using dielectric materials in general. In none of the above scenarios is the observed shape of the localised maximum at the antisolar point easy to explain.

Before leaving this sketch of the state ZL knowledge in 1970-4, when my experiment took place, it is interesting to note that the era of space-borne observations of the interplanetary medium was just beginning, and at this stage, things were not going as well as might have been hoped! Sparrow and Ney (1973), based on data from OSO-5, concluded that there was essentially no variation in the brightness or the polarisation of the ZL from 1969 to 1973. Levasseur and Blamont (1973), on the other hand, using data from satellite OS2 "Tournasol" disagreed entirely. It might be interesting (but outside the scope of this work) to revisit these two papers in the light of what we now know from the 1984 IRAS revelations. This, coupled with the similarly inconclusive results of particle collection mentioned earlier, set a tone of uncertainty in the results of *in situ* studies, which around 1970 gave good reason

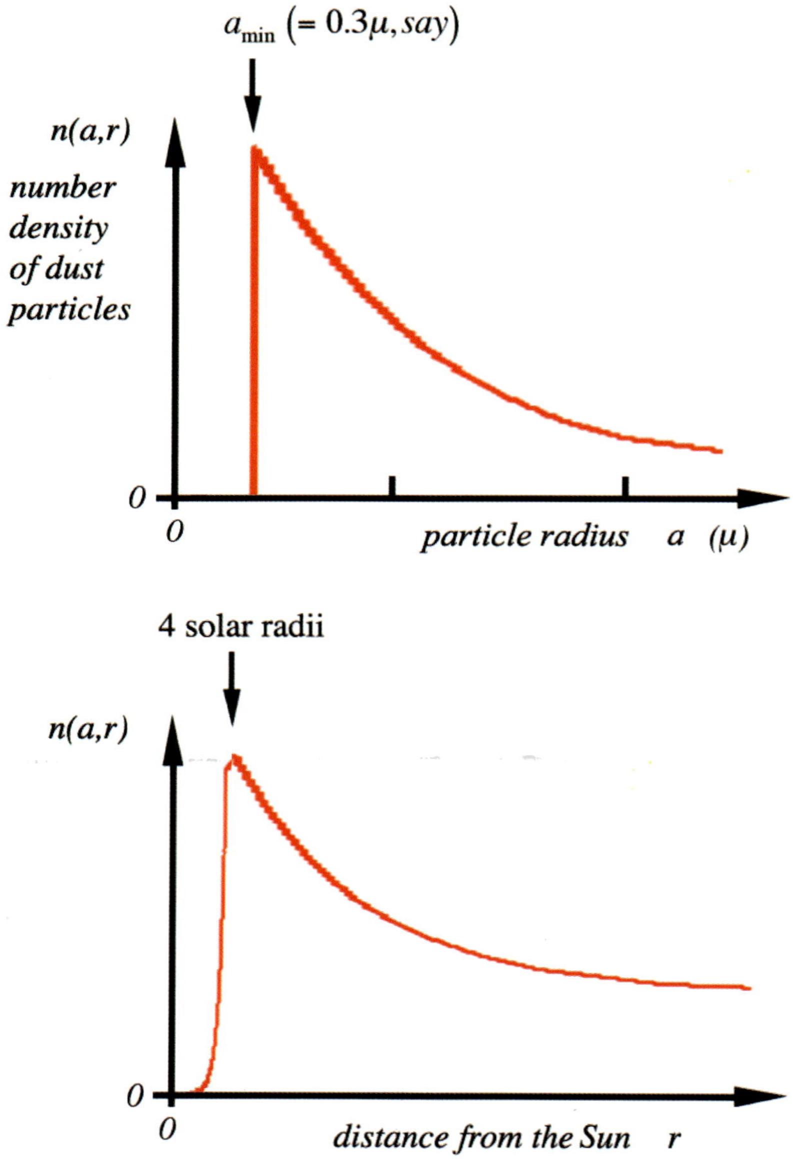

Figure 1.4 A simple model of Zodiacal dust density as a function of particle radius and distance from the Sun.

to persist in Earth-based research on the ZL, in particular with the relatively new technique of *Doppler radial velocity measurements*.

1.2.5 Doppler investigations of the ZL prior to 1970

The approach of radial velocity measurement is especially valuable in the study of the ZL, since it provides completely independent estimates of some of the parameters – *viz*. speed and direction distributions in the cloud, leading to insights into its formation and evolution.

Christian Doppler, an Austrian mathematician and physicist working in Prague, was the first to draw attention to the change of the pitch of sound waves due to a relative motion of the emitter and the observer, an effect very noticeable in the siren noise on a passing police car, and thereafter known as the Doppler effect. He was able to offer an explanation, in terms of the way the wavelengths perceived by the observer were 'squeezed or stretched' by the relative motion, and proposed, in his paper in 1842, translated as *"On the Coloured Light of Double Stars and Certain Other Stars of the Heavens"*, that a similar effect would be seen with light waves. It is easy to visualise that light waves emitted from a moving source are squeezed closer together (as perceived by us) if the source is approaching us, producing a shift in wavelength to the blue, and stretched out if the source is moving away, so that we see a red shift. In 1848, French astronomer Hippolyte Fizeau, the first to make a serious estimate of the speed of light, confirmed experimentally that the Doppler effect changed the positions of spectral lines, by amounts proportional to a source's radial velocity relative to us.

The Doppler shift $\Delta\lambda$ from a source of light of wavelength λ, moving at a radial velocity V relative to us, is given by the simple relation

$$\frac{\Delta\lambda}{\lambda} = \frac{V}{c} \qquad \text{where } c \text{ is the speed of light.}$$

In 1862, Huggins and Miller remarked that the analysis of Fraunhofer lines in stars could lead, not only to analysis of their material compositions, but also to estimates of their motions. *"If the stars were moving towards or from the Earth, their motion, compounded with the Earth's motion, would alter to an observer on the Earth the refrangibility of the light emitted by them, and consequently the lines of terrestrial substances would no longer coincide in position in the spectrum with the dark lines produced by the absorption of the vapours of the same substances existing in the stars."* Thus was born the science of spectral radial velocity measurements.

We can note in passing that the Doppler effect became a pillar of the evidence for the *'Expanding Universe'* as we perceive it today. The above relationship was invoked by Vesto Slipher in 1917, who found that the light from the 21 of the 25 galaxies he studied showed a red shift, indicating that they were receding from us, and in 1929 Edwin Hubble was able to show that the speed of their recession was proportional to their distance from us.

Much closer to home, we have seen that Blackwell and Ingham (1961) used the concept of 'smearing' of absorption lines due to high Doppler shifts to place an upper limit on free electrons in the interplanetary medium. This introduced Doppler techniques to the study of the Zodiacal Light, and it was Ingham who laid the foundations for use of the Doppler method for investigations of the motions of interplanetary dust.

The profile of an absorption feature in the ZL spectrum in a particular area of sky corresponds to a unique line of sight through the Zodiacal Cloud, and is a combination of the effects of all the light-scattering dust particles along this line. We can regard the spectrum as the summation of contributions from all the elements, or 'cells' in the line of sight; each has a radial velocity relative to the Earth which is a combination of its own motion and that of the Earth, and leaves its 'signature' on the resulting combined spectrum. By comparing such a theoretical prediction with experiment, for various different directions relative to the Sun, information can be derived on relative number and size distributions *and* velocity distribution in the cloud, quite independently of other observations. Typical geometry applying to a radial velocity element, or 'cell', is shown in Figure 1.5.

This view from 'above' the plane of the ecliptic shows the geometry for a cell of dust, viewed from the Earth, along a line of sight at an elongation angle ε from the Sun. In the particular case illustrated, the dust is moving with speed V_c, in a circular solar orbit, and therefore has no radial velocity relative to the Sun (in fact the orbits may not be circular, but may be nearly so). The cell 'sees' the continuous visible light emission spectrum of the Sun, crossed by dark Fraunhofer lines of absorption, with zero shift. Dust in the cell scatters light towards the Earth. But this cell is moving away from the Earth radially with a component of its velocity along the line of sight $V_c \sin\theta$, while the Earth's orbital motion, velocity V_E, produces an opposing, closing component of velocity along the line of sight $V_E \sin\varepsilon$. The resulting radial velocity to be used in applying the Doppler shift is thus $V_c \sin\theta$ - $V_E \sin\varepsilon$. We can eliminate the cell's velocity as follows. We know that for circular orbits:

$$\frac{V_c^2}{V_E^2} = \frac{R}{r},$$

where r is the distance of the dust cell from the Sun, and R is the distance of the Earth from the Sun. In the Earth-Sun-Cell triangle the Sine Law gives:

$$\frac{r}{\sin\varepsilon} = \frac{R}{\sin\theta}, \quad \text{or} \quad \frac{R}{r} = \frac{\sin\theta}{\sin\varepsilon}$$

hence $$\frac{V_c}{V_E} = \sqrt{\frac{\sin\theta}{\sin\varepsilon}}$$

And the required relative radial velocity ΔV is given by

$$\Delta V = V_E \sqrt{\frac{\sin\theta}{\sin\varepsilon}} \sin\theta - V_E \sin\varepsilon .$$

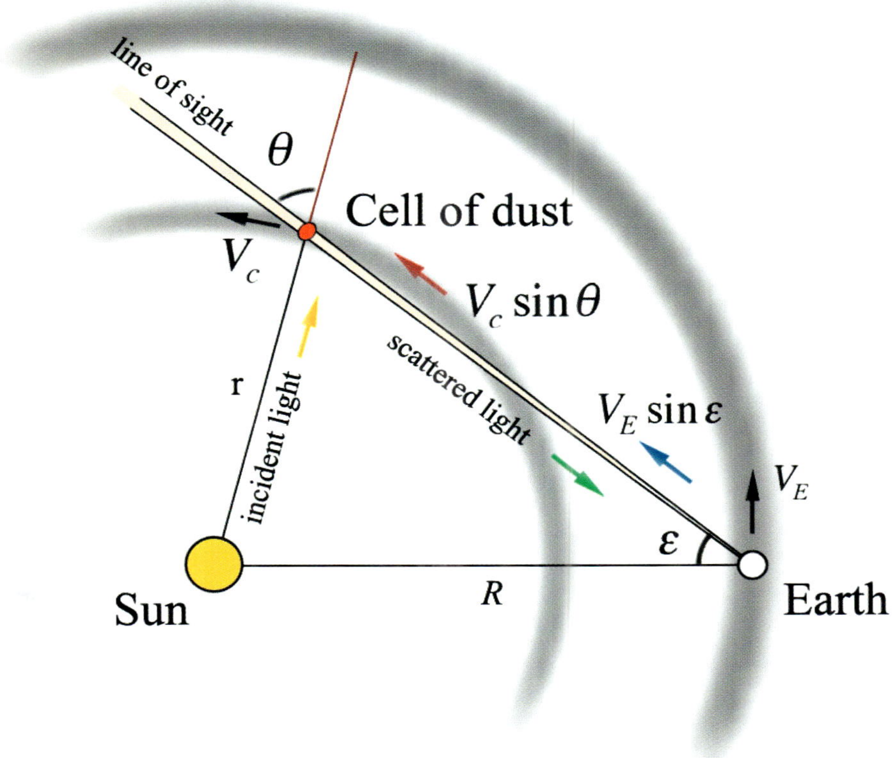

Figure 1.5 Illustrating a line-of-sight integration for prediction of a Doppler-shifted absorption line spectrum from a rotating dust cloud model.

The resulting contribution of this cell to the spectrum of the ZL is light with a red shift of $\Delta\lambda$, where:

$$\Delta\lambda = \frac{\lambda}{c} V_E \left(\sqrt{\frac{\sin\theta}{\sin\varepsilon}} \sin\theta - \sin\varepsilon \right)$$

Ingham first arrived at this formula in his paper (1969) in a slightly different form, using cosecs.

An integral of this expression for all cells along the line of sight, weighted according to the predicted density of dust for each cell position, and the percentage of incident light re-directed towards the observer, gives an 'average' radial velocity, and hence an average wavelength shift for any absorption line, which can be compared with the shifts found by experiment. Clearly this is an approximation; a full treatment integrates the intensity at each wavelength, and predicts the profile of the absorption line in detail.

In 1963, Ingham (1963) published his theoretical predictions, calculated in this way, of the manner in which the shape of a solar Fraunhofer line would be modified by scattering from a rotating dust cloud medium (Figure 1.6), in the ecliptic plane, as a function of elongation from the Sun. He showed that

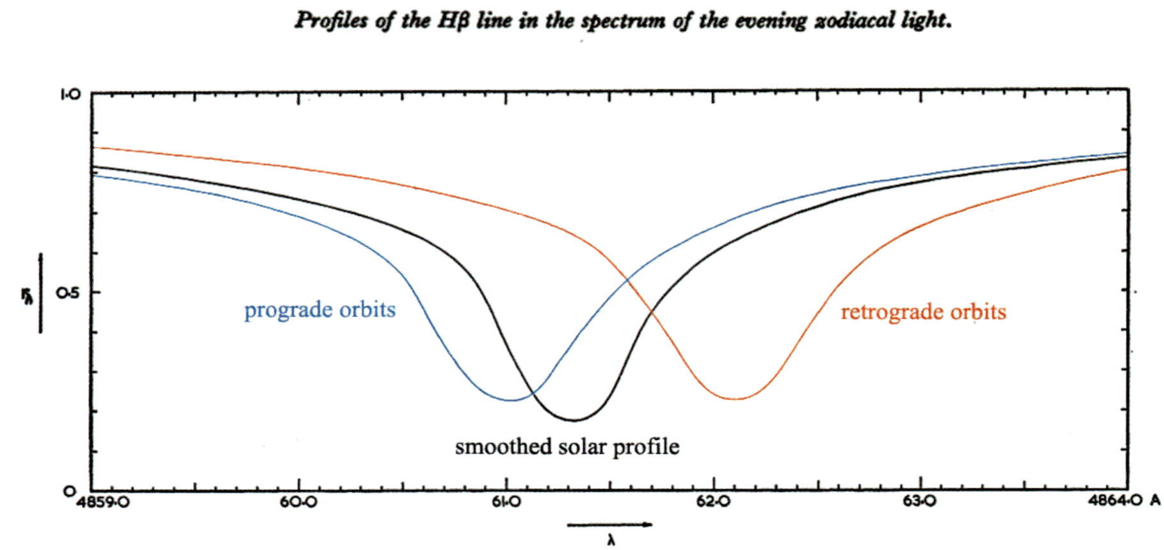

Figure 1.6 Ingham's (1963) theoretical predictions for the shape of a modified Fraunhofer line (elongation 30°) in sunlight scattered by a rotating circum-solar dust cloud.

a broadening and overall shift of the wavelength of the line ought to be detectable, giving information with which to map the motions of Zodiacal dust. All that was now needed was the observational evidence to compare with this theory.

Attempts to record a high-resolution spectrum of the ZL as a tool for revealing and discriminating between motional signatures, are limited only by the resolution-luminosity (R-L) product of the instrument and time available to spend on observing each point. The need for high R-L product was a good reason for choosing an interferometric spectrometer, the Fabry-Perot etalon (Jacquinot 1954), but as we will see, there are other attractive alternatives.

Beginning in June-August 1962, under the leadership of Professor James Ring at Chacaltaya in Bolivia, Hindle, Reay and Ring (1968) employed a Fabry-Perot etalon in series with an Ebert Spectrometer, with a resulting resolved width of about 1.2 Angstroms (Å), to record Zodiacal Light spectra in the region of the H_β absorption line over a range of about 10 Å, with the intention of determining Doppler shifts in this line. Detection was by cooled photomultiplier and counting electronics, a precedent followed in many subsequent experiments of this type. Instead of the absorption line they had hoped to see, they found strong evidence of H_β emission. Only a small number of observations were made, mostly of the evening ZL, and the conclusions were:

(a) That the profile of the Balmer H_β line in the spectrum of the Zodiacal Light did not match that of the Sun, either in position or shape.
(b) That some scans showed clear evidence of emission at the H_β wavelength, especially near the Galactic plane, and it was proposed that this emission was from Hydrogen in the Galactic plane.

In 1964, Prof. Ring's team made a new set of observations, using a 3" Fabry-Perot and an interference filter premonochromator (Daehler *et al* 1968). The width of the pass-band of the system was 1.5 Å, and about 50 scans were obtained of morning and evening ZL. The most immediate result was the confirmation of H_β emission, which still obscured the absorption line. The emission was stronger in the morning than in the evening, and apparently had a maximum intensity at 70 degrees elongation from the Sun in the ZL. So strong was the emission that it was difficult for these observers to draw any conclusions about the obscured absorption line shape.

In November and December 1967, Reay and Ring repeated their observations from a site at Testa Grigia in the Italian Alps, looking at regions of the sky clear of the ecliptic, in both morning and evening, with an improved instrument. They found more H_β emission, and not only in the plane of the Milky Way. An automatic stepping system and careful calibration were introduced so the identification of the H_β wavelength was now beyond doubt. Even in a moonlit sky (contributing, of course, reflected sunlight with an unshifted H_β line in absorption) the emission was strongly seen. Shifts of the line were observed in some scans in the evening Zodiacal cone – all towards the blue. This was still far from the simple picture of a shifted absorption line which the experimenters had looked for.

This time, however, Reay and Ring attempted to subtract the emission lines from the spectra, to reveal an interpolated absorption feature which hinted at the shifts they were looking for. These correlated tolerably well with a prograde dust model, but the probable errors were high, and the data was confined to evening ZL elongations between 40 and 120 degrees. Nevertheless these data (plotted as green circles in Figure 4.1) stood for some years as the results with which motional models were compared. In their letter to Nature in 1968, Reay and Ring announced their intention to overcome the problems of obscuration of the absorption feature by H_β emission, by switching their attentions to the MgI line at 5183.6 Å.

Returning to the absorption line data, Bandermann and Wolstencroft (1969) attempted a better fit to the Reay and Ring data, by basing their model on elliptical dust orbits, rather than circular, but the differences in the predictions were small, and Reay (1969) produced, theoretically, larger variations by including the effects of radiation pressure on orbital speeds. These effects are large in the case of sub-micron particles. Other attempts were made to construct the theoretical shape of an absorption feature from rotating dust-cloud models, notably James (1969).

Vanysek and Harwit (1970) pointed out that the Reay and Ring data were actually not inconsistent with the theory that a 'Cis-Terrestrial Cloud' (CTC) was responsible for the ZL, and predicted a drop in the wavelength shift at elongations of about 70 degrees. They also showed that the application of full Mie Theory to the scattering problem produces predictions very similar to the Hulst-Ingham-Reay method of treating reflection and refraction separately (Ingham 1961).

New experimental data was obtained by James and Smeethe (1970), using the new target absorption line, MgI (5183.6 Å), who attempted to make observations of the more elusive (because of twilight and atmospheric extinction) lower elongation regions in the both the morning and evening ZL. Their results (shown as brown crosses in Figure 4.1) did show high shifts at low elongations, which is what the circumsolar prograde models predict, but the radial velocities they found looked too large to lie on the $p = 5$ theoretical curve (see Chapter 4), and again the probable errors made it impossible to support any firm conclusions. The observations were limited to a very small range of elongations, so no significant curve-fitting could be done to discriminate between models. James and Smeethe concluded that their observations were consistent with a dust distribution model in which $p < 3$, and most of the dust was in prograde solar orbits, but the matter remained unproven.

Throughout these studies prior to 1970, it had been hoped that the radial velocity patterns found would discriminate between dust in prograde and retrograde orbits around the Sun: if the dust were determined to be moving in orbits around the Sun in the same sense as the Earth ('prograde'), an asteroidal origin would be favoured. In this hypothesis, due to Fesenkov (1958) collisions and/or cosmic ray bombardment within the asteroid belt produce debris, which slowly spirals in towards the Sun. Other authors of the period, notably Whipple, favoured cometary debris as the principal component of Zodiacal Dust, an idea first suggested by Fesenkov in 1914. Since many comets (in fact perhaps a majority of those with long periods) have retrograde orbits, presumably the discovery of any retrograde elements must favour a cometary component. It is interesting that these two opposing views of the origins of interplanetary dust are still hotly defended in 2007 (see Chapter 4). A Zodiacal Cloud origin *coeval* with the Earth and other planets was thought by most to be unlikely in 1974, since the gradual collapse of the orbits of small particles, and their probable eventual vaporisation by the Sun (Dobrovolskii and Egibov 1973), over a timescale short relative to the age of the Solar System, requires that the material is continuously replenished. If the source of replenishment of the material is outside the Solar System, we might expect an entirely different pattern of velocities. However, opinions on the relative strengths of the various forces acting on a particle vary greatly, so the question of the age of the Zodiacal Dust may be regarded as still unresolved, and certainly was in 1970. If the dust did turn out to be a remnant of the creation of the Sun's whole planetary system, we would still expect a predominantly prograde cloud.

To sum up, in 1970, uncertainties in observations of radial velocities, combined with uncertainties as to which theoretical model should be used as a comparison, left the question of the real motions of Zodiacal Dust completely open.

1.3 Review of the period of my observations of the ZL, 1970–1974

It was the intention of this PhD study to fill in the gaps and uncertainties left in observational data in 1970; to make a comprehensive survey of a Doppler-modified Fraunhofer line in the ecliptic plane at all elongations from the Sun, and hence constrain an accurate model of the Zodiacal Cloud's configuration and velocity structure. The spectral line chosen was the MgI 5183.6Å line, because of its (hoped) freedom from emission line contamination, and a new pressure-scanned Fabry-Perot interferometer was designed and built to make the observations.

In 1972 and 1974, we, Hicks, May and Reay of Imperial College Infrared Astronomy Department, under the supervision of Prof. James Ring, published, in two papers, the results from the research which forms the basis of this thesis. The first paper in Nature (Hicks, May, Reay 1972) reported the MgI emission which had unexpectedly turned up in the new spectra. The second paper in MNRAS (Hicks, May, Reay 1974) reported the evidence, found in our Zodiacal Light spectra, of radial velocity shifts in the MgI absorption line, the first comprehensive mapping of such shifts ever achieved. The paper confirmed for the first time with any certainty that the pattern of Doppler shifts was consistent with a high proportion of circum-solar dust moving in prograde orbits, though some significant asymmetries between morning and evening velocities were noted. A third paper detailing further interpretation with reference to dust cloud models, including a putative interstellar component, was never completed. The data obtained from this project, along with some comparisons with simple models of the ZDC, are detailed in Chapter 4, which constitutes the first full publication of the results of my research 1970-4.

1.4 Review of recent developments in ZL research, 1974–2007

1.4.1 Overview

Returning to the field of Zodiacal Light research after 30 years, I found I was in a good position to obtain a perspective on what progress had been made in the intervening time. I am indebted to four excellent papers for much of the information in this review – Grogan *et al* (2001), Reach *et al* (2003), Kelsall *et al* (1998), and Sykes *et al* (2004). My first observation was that, in keeping with the acronymic style of pigeon-holing now so universal, in which everything from espionage organisations, to diseases, to furniture warehouses, find themselves summed up in three letters, interplanetary dust particles are now referred to as "IDP"s. And the business of detecting them, and their composition, and orbital elements, has progressed enormously; we even have some in our hands, to study on Earth.

Secondly, it soon became apparent that there are now historically two distinct eras within modern ZL research: pre-IRAS and post-IRAS (InfraRed Astronomy Satellite). In the period 1974-1982, observations were extended, and the existing dynamical models updated, but the overall picture remained essentially unchanged, except for significant advances in *in-situ* collection techniques, and the analysis of meteoric impacts on Lunar samples (Singer 1969). It was the data gathered from the IRAS satellite in 1983 (while my rock group Queen was touring football stadia in South America) which finally moved our picture of the Zodiacal Cloud on, in a giant step, from diffuse oblate spheroid, the limit of anyone's imagination since the 17th century, to a structured, dynamic and complex dust ensemble.

This features the now well-known ‘IRAS Bands’ (Low *et al* 1984), (see Figure 1.7), of which at least 4 pairs are now known, contributing about 10 per cent of the total Zodiacal emission in the infrared, and, on a small scale, cometary dust trails (around 8 of these have been seen at the time of writing). Moreover, it revealed a local dust cloud of a kind near the Earth, but the idea of a GDC was, at a stroke, replaced by a local toroidal dust intensification near the Earth’s in its path, due to resonance in the dust orbits; effectively a ‘dust lane’ stretching out behind the Earth in its journey around the Sun. This research also revealed (Hauser *et al* 1984) that the axis, the line of brightest scattering of the ZL, changes its orientation to the ecliptic plane, depending on where the Earth is in its orbit. The position of the axis varies in a yearly sinusoidal cycle with a maximum tilt of about two degrees, bringing about an expected seasonal intensity variation of around 15 per cent, in addition to a 5 per cent variation in a different phase, resulting from the offset of the Sun from the centre of the cloud. This, strangely, brings to mind once more the findings of the patient Rev. Jones in 1856, who did see a variation over the two

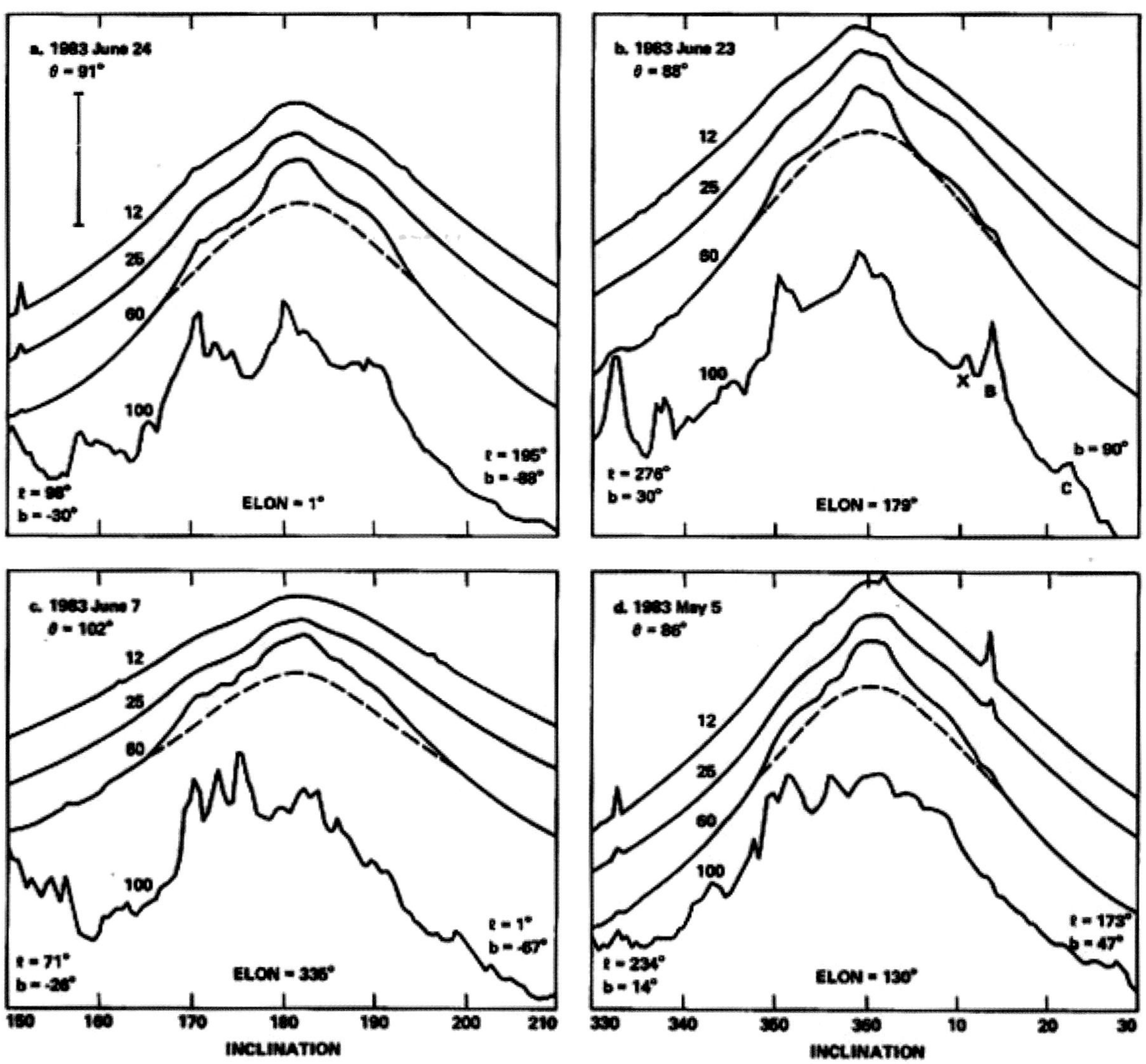

Fig 1.7 Low *et al’s* discovery of the Zodiacal Dust Bands. For details see Low *et al* (1984).

years in which he made his naked-eye observations of the contours of the ZL as he saw them from his boat. He saw what he thought were parallax-induced variations in the position of the axis, but since his voyage moved him not only in latitude, but around the Sun too, over a period of two years, it is perhaps possible that his maps record the first documentation of the inherent asymmetry of the ZC with respect to the ecliptic plane.

Dermott *et al* (1992) achieved an impressive theoretical fit to the yearly cyclic variation, using their own updated 'SIMUL' model of asteroidal dust of radius 3.4 μm, spiralling slowly inwards under the influence of Poynting-Robertson drag (Robertson 1937), and perturbation of the planets. A little later, Leisawitz *et al* (1994) using the new data from IRAS, but also that from the DIRBE survey aboard the COBE satellite, extended the analysis to link the dust-band data with particles of varying sizes produced by collisions of asteroids in the Eos family. All the new analyses led to a much more detailed picture: a finely tuned version of the 'traditional' lenticular-shaped ZC, now symmetrically distributed not about the ecliptic, but relative to a plane dictated by planetary perturbation, inclined at about 2 degrees to the ecliptic, and centred on a point 0.13 AU from the centre of the Sun. This updated view of the 'traditional' ZC was theoretically modelled with a combination of cometary and asteroidal contributions, the proportions of which are still under debate. But the new picture also included a three-dimensional notional map of the newly discovered, relatively stable 'bands' (attributed to toroidal concentrations of material), oriented at roughly one degree angles to the plane of the lenticular cloud, identified by Low *et al* (1984) and studied in detail by Grogan *et al* (2001) with matter spiralling in, under the influence of Poynting-Robertson drag, from collisions in certain families of asteroids. Added to this are the discrete trails of dust left by comets, seen unmistakeably for the first time in the IRAS composite images, (Sykes *et al* 1986) and a circumsolar ring at just outside 1 AU, as predicted by Gold, in 1975, observationally supported by Dermott *et al* (1994), and confirmed by Reach *et al* in 1995. The IRAS observations showed a consistent 3 per cent increase in intensity at all its wavelengths behind the Earth, looking directly at its line of motion, relative to that seen in front of the Earth in its orbit. This is consistent with the predictions for a resonant ring of width 0.4 AU with a cavity in front of the Earth. Dermott *et al* noted that this must produce an increased rate of capture in the Earth's upper atmosphere, and that a seasonal variation ought to be seen, according to where the Earth lies in relation to the Zodiacal Cloud. They also saw this as a possible connection to the accretion of Life-bearing carbonaceous particles in the history of the Earth. Those particles not captured by the planets, of course, suffer a different fate: under the influence of Poynting-Robertson drag, and solar wind braking effects, the Zodiacal Cloud loses one ton of material per second, as the dust spirals ever inwards, to be vaporised at a distance of a few solar radii from the Sun.

The researches of these teams, as noted by Kelsall *et al* (1998), were all initially directed towards evaluations of infrared background emissions, from Galactic and extragalactic sources. It was a need to eliminate the foreground 'contamination' by the local interplanetary dust which led to their careful analysis of the ZC. Rowan-Robinson *et al* (1990) derived, from early IRAS data, a model of 12-100 micrometre emission by the Zodiacal Cloud, which I will refer to hereinafter as the RHVW model, which they were later able to use to subtract from the observational maps, to show the 'shoulders' of the IRAS bands (see details in 1.4.2.1).

So the Zodiacal Light can certainly no longer be thought of as smoothly symmetrical about the ecliptic plane. IRAS's data collection was in the infrared, at 12, 25, 60 and 100 μm, and all these observations are of emission, rather than scattering of sunlight, but there are now claims, from Ishiguru *et al* (1999), that corresponding features due to the existence of the IRAS bands exist in scattered light at visible wavelengths. They scanned the surface brightness of the ZL at right angles to the ecliptic plane, and discerned the 'shoulders' of discrete bands, superposed on the smooth curve of brightness associated with the 'traditional' lenticular ZC. This research begs the interesting question of why 50 years of careful photometry of the ZL from the Earth never revealed these 'shoulders' before. Perhaps this is another example of LFS – the Lucky Find Syndrome – nobody saw the 'shoulders' because nobody thought they should be there. Ishiguru *et al* are also one of the few modern observing teams to follow up on the Gegenschein, and claim to have detected parallax in the optical bands in the Gegenschein area, placing the source material at 1.6 AU from the Sun. In a recent collection of accounts of progress in ZL observations, the book *"Interplanetary Dust"* (Grün *et al* 2001), the Gegenschein receives only four mentions, three in an (excellent) historical review, and one in the glossary.

At this point, since the newly discovered detailed structures of the ZL must in future play an essential role in constructing models of radial velocity patterns, it is useful to look at the important recent advances in more detail.

1.4.2 Infrared Surveys

1.4.2.1 Modelling based on IRAS data

Rowan-Robinson *et al* (1990) investigated the distribution of Zodiacal Dust as a function of position out of the plane of symmetry. Following Leinert (1985) and others, they looked for a fit to the IRAS flux measurements, using a functional dependence of number density of the dust grains:

$$n(r) = n_0 \left(\frac{r}{A}\right)^{-\gamma} f(\beta_0),$$

where n_0 is the number density near the Earth's orbit, A = 1.AU, β_0 is the elevation from the symmetry plane of the dust, and $\gamma \approx 1.3$ from Helios data (Leinert *et al* 1981). A popular form of the function $f(\beta_0)$ was the 'Fan' model:

$$f(\beta_0) = \exp(-\alpha \sin|\beta_0|),$$

and Leinert *et al* (1981) put $\alpha = 2.1$.

Rowan-Robinson *et al* in this paper explored models with a modified form

$$f(\beta_0) = (\cos\beta_0)^Q \exp(-\alpha \sin|\beta_0|).$$

A particular form of this with $Q = 1$, $\alpha = 2.6$ had been used by Murdock and Price (1985) to fit the 'ZIP' data (from Aries rockets).

The cloud was assumed to be made up of isotropically scattering 'grey' particles (for which they need to be of radius greater than a millimetre), and a temperature was assumed of the function:

$$T(r) = T_0 \left(\frac{r}{A} \right)^{-1/2}$$

No attempt was made to model for the IRAS dust bands at this point, or for the fact that the Sun might not lie at the centre of the symmetry plane of the disc, which had been reported by Dermott *et al* (1987); but provision was made for the symmetry plane of the dust not to be coincident with the ecliptic plane, and to be inclined to it at an angle *i*, with a longitude of the ascending mode Ω .

The best fit was found to be: $\Omega = 79$, $i = 1.1$, $Q = 0$

and a figure very different from Leinert's 2.1 estimate, of $\alpha = 4.9$. This means physically that in this model, the dust is closely confined in the direction perpendicular to the plane of symmetry, and not so closely confined in the direction of radial distance from the Sun; the disc is quite flattened.

Rowan-Robinson *et al* (1991), in a following paper, used this model to subtract the main, smooth distribution of this theoretical cloud, revealing the shape of the dust bands. Figure 1.8 shows the result, at $60\mu m$. The contours are $0-5\ MJy\ sr^{-1}$, in steps of $1\ MJy\ sr^{-1}$.

These authors concluded that there was a dust band at $\beta = \pm 10$, at colour temperature about 170°K, and 'warmer' bands at $\beta = \pm 5$, barely detected at $60\mu m$, with a colour temperature of 250°K. The inner band was identified by Dermott *et al* (1985) as being due to debris from collisions in the Themis asteroid family, situated at about 3 AU. It is proposed that the collisional dust due to this 'family' stays in the vicinity of the asteroids themselves, forming a diffuse toroid of material, which only slowly spirals in towards the Sun under the influence of drag from the Poynting-Robertson effect and solar wind. Members of families of asteroids all have the same orbital eccentricity, and the debris, at least initially, preserves this. Why are the bands in pairs? Dermott and Nicholson (1989), in a numerical study, suggested that the toroidal distribution formed by collisions has a dust density *"higher near the surface of the torus than within it"*. Thus, looking at a sectional view through the torus, we see a

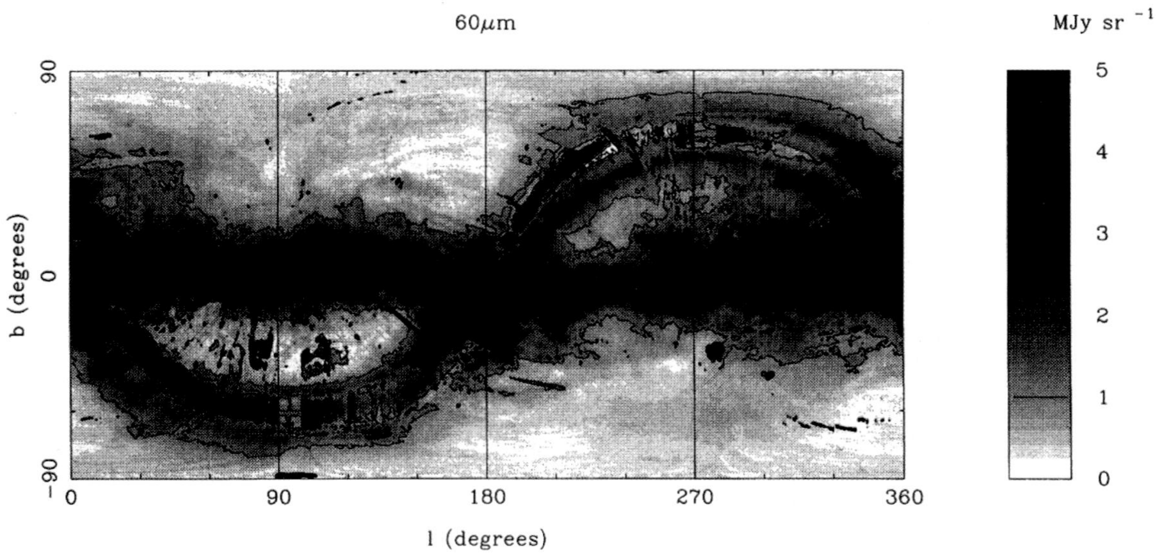

Figure 1.8 From Rowan-Robinson *et al* (1991). Grey-scale and contour map of the 60 micron emission, in Galactic coordinates, after subtraction of RHVW and application of a correction factor.

concentration of emission at the edges. In the case of the Themis band, therefore, the torus is 5° thick from our viewpoint, or in actual size, 0.25 AU. The outer bands were later identified as being associated with the Eos family of asteroids, which also reside at about 3 AU, but whose orbits are presumably more scattered in their inclinations to the ecliptic. In addition it was noted that there could be a component of interstellar dust, forming a disc located at 50 AU, in the 'inner' Oort Cloud, beyond the orbit of Pluto, which is thought to be the origin of short period comets.

Jones and Rowan-Robinson (1993), noting that the previous model did not make good predictions at low ecliptic latitudes, and also did not account for the IRAS Dust Bands, modified the 'Fan' empirical model at low values of z, where z is the perpendicular distance of the dust particle from the plane of symmetry, and thus $\sin\beta_0 = z/r$. A new model was constructed, using components corresponding to an inner smooth fan of relatively high density, an outer fan, discontinuously beginning at about 1.5 AU, and superposed toroidal distributions corresponding to Themis and Eos. Based on the hypothesis that 'old' drag-degraded asteroidal Bands are at least significant contributors to the smooth cloud, the Fan model was assumed to contain essentially no material beyond a distance of 3 AU from the Sun. The major source of perturbation to dust grain orbits was assumed to be the planets. For areas close to the Sun, the parameter γ was set to 1, since this is what the Poynting-Robertson effect indicated might be the steady-state situation, not 1.3 as used in analysing the Helios data. In this paper, the authors now modelled the density profiles of the bands themselves, using an analytical model for two families of asteroids, Eos and Themis. They also looked at modelling the density profiles of the *broad* bands seen in Figure 1.8, which the authors suggest may be due to the 'scattering out' of smaller particles from the narrow bands. The model they used is an adaptation of the narrow band model, but with two exponents instead of one to simulate the broad 'plateau' in the emission profile. The authors comment that it was this broad band which was actually first discovered by Low *et al* (1984). The best fit for this broad band model was added in to the fans and the narrow bands, plus an isotropic component, to achieve an overall best fit to the thermal emission mapping. In the fitting of the new Fan model, they found these new values for the variables:

$\Omega = 76, \quad i = 1.3, \quad \alpha = 8.0, \quad Q = 2.85$.

Figure 1.9 shows the appearance of the modified combination 'Fan and Band' model.

Jones and Rowan-Robinson in this analysis concluded also that the asteroid belt only accounted for about a quarter of the dust in the main belt. The rest was presumed due to comets, and material from the Oort Cloud, plus the possibility of an interstellar component.

Kelsall *et al* (1998), modelling the COBE data, referred to Dermott *et al* (1996) and Leinert *et al* (1998), discussed the evidence from the capture of particles, and concluded that 45% of the interplanetary dust was asteroidal. This may be compared with Durda and Dermott (1997) who estimated that 34% was from collisional destruction of asteroids – 10% from family members, and 24% from the main band. So it would seem from these analyses that perhaps one third of all interstellar dust is from asteroids. We have seen that Baguhl *et al* (1996) found evidence of interstellar component, but Kelsall estimated, from their own fitted model, that its contribution is 'small' (less than 15-20%). The distribution model of Kelsall *et al* (1998) is almost exactly along the lines of Rowan-Robinson *et al* (1990) except for the use of different symbols. Their formula for $n(r)$ in terms of a function of z_c / R_c corresponds to $f(\beta_0)$. The parameters they arrive at for best fit are:

$\gamma(or\ Q) = 0.94, \quad \beta = 4.14, \quad \Omega = 77.7, \quad i = 2.03.$

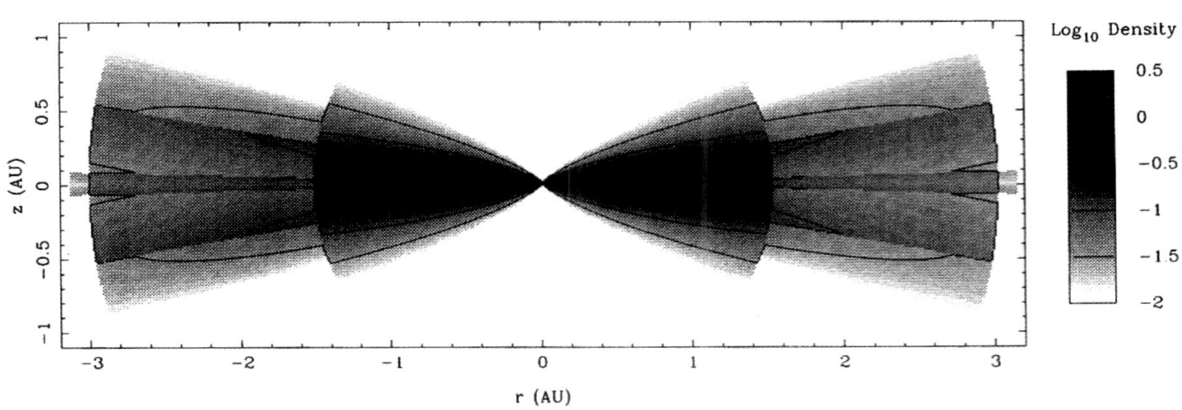

Figure 1.9 From Jones and Rowan-Robinson (1993). A density map of the best-fitting combined Fan and Band model. The grain density is normalised.

This paper gives a very useful picture of density contours in a slice across the plane of symmetry (Figure 1.10), showing the appearance of dust bands and the Earth Ring.

Grogan, Dermott and Durda (2001) were the first to construct a model of the ZC dust bands including a significant population of large particles, as suggested by the observations of Love and Brownlee (1993). Moving away from empirical models entirely, Grogan *et al* followed the journey of asteroidal debris through the Solar System, constructing a model based physically on individual dust particle orbits, taking into account the size-frequency distribution of the material. A given asteroid collision produces debris according to some size-frequency distribution, given by

$$n(D) \propto D^{2-3q},$$

and, for 'collisional equilibrium', Grogan *et al* put $q \approx 1.8$.

They noted that the average inclination of the Eos family of asteroids did not exactly match that of the 'ten degree' (actually 9.35 degree) dust band, but, working from the break-up of a single family member in the lower inclination end of the family, they were able to produce an impressive fit, shown in Figure 1.11.

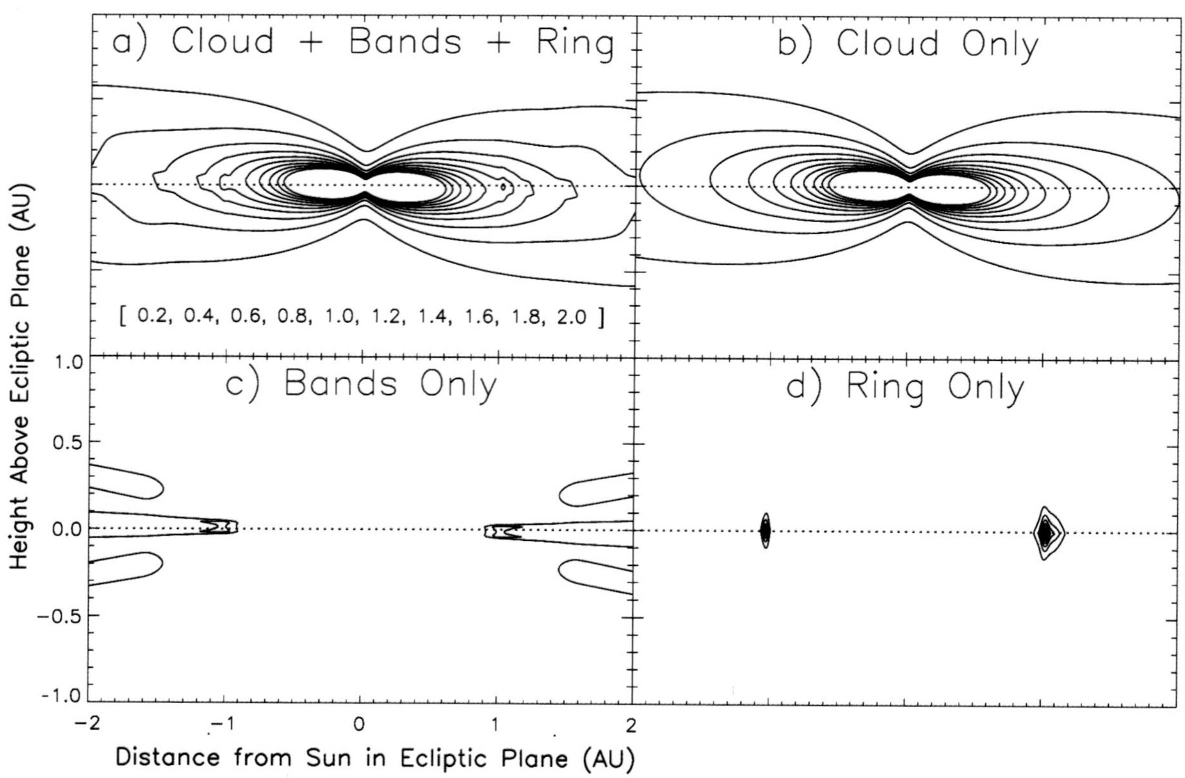

Figure 1.10 From Kelsall *et al* (1998). Isodensity contours of the IPD model components, shown for a cross-sectional slice perpendicular to the ecliptic plane: (a) all components combined; (b) smooth cloud; (c) dust bands; (d) circumsolar ring. Contour levels in (c) and (d) are a factor of 8 smaller.

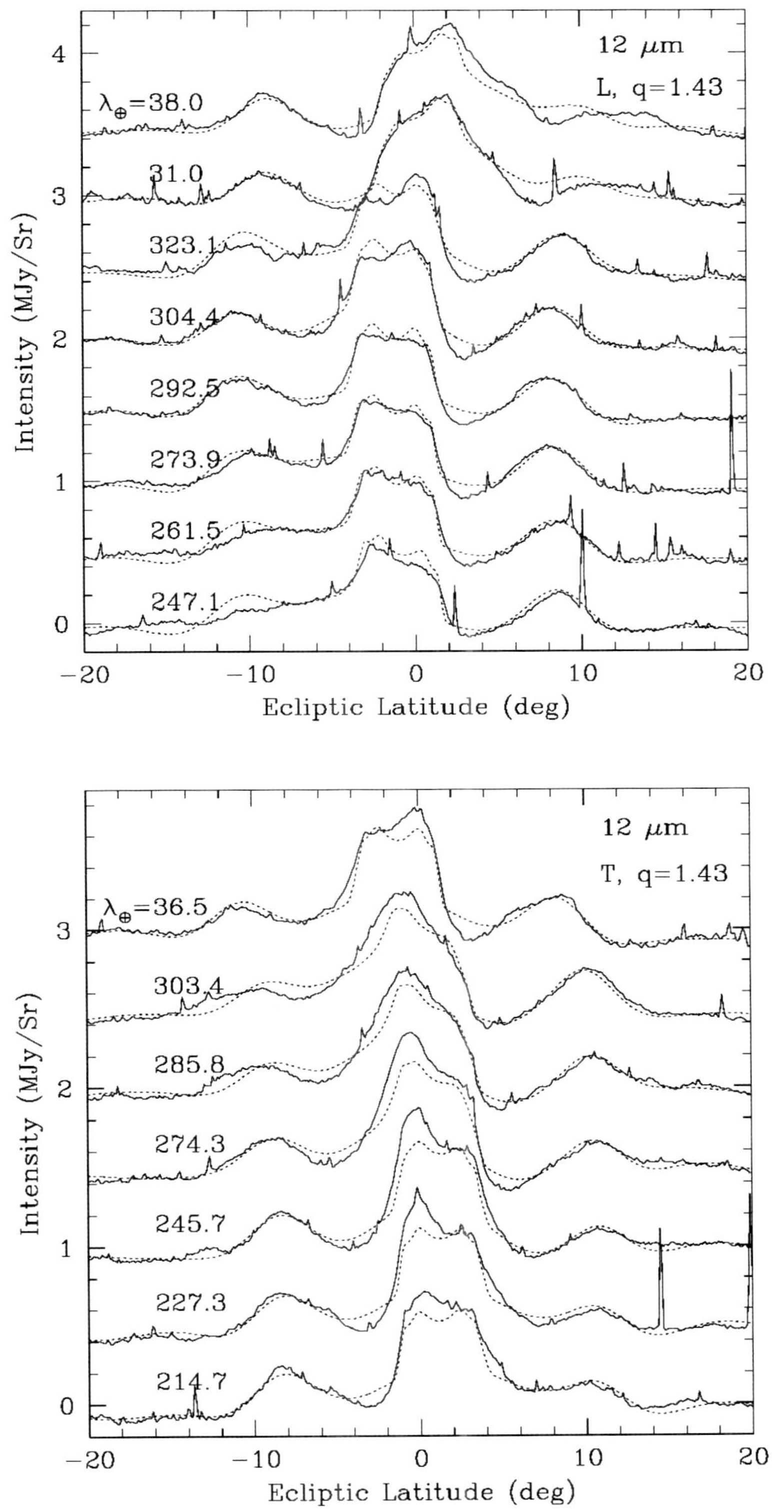

Figure 1.11 Best-fit dust band model (dotted curves) of Grogan, Dermott and Durda (2001) to the 12 micron IRAS observations in both the leading (top) and trailing (bottom) directions of view. This model has a size-frequency index of 1.43, indicating that large particles dominate the distribution.

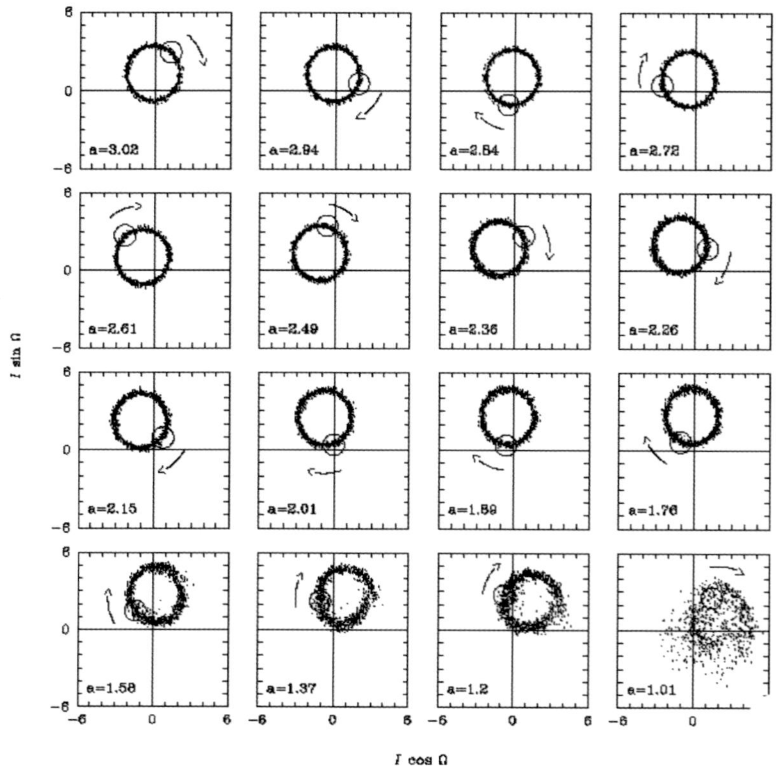

Figure 1.12 From Grogan, Dermott and Durda (2001). Showing the evolution of the inclination distribution of a wave of 10 micron diameter Koronis dust particles migrating from the asteroid belt toward the Sun (Kortenkamp and Dermott 1998). The particles were released 52,000 years ago with a mean semi-major axis of 3.02 AU. The wave has now reached 1.01 AU, indicating approximately 3,500 years between each frame. The motion of the 'centre of mass' of the distribution represents the changing I_f and Ω of the wave. The radius of the circular distribution remains constant until the wave approaches Earth and many particles are gravitationally scattered. The large open circle in each frame follows the evolution of a single orbit and reveals the clockwise precession of Ω_p.

The excess velocities of collision products will be much less than the orbital speed of the parent particles (15-20 km/s), so the orbits of the individual fragments will be similar, their orbital elements only slightly perturbed in the beginning. But in only 100 to 1,000 years, this small differential in relative velocity (10-100 m/s), corresponding to a minor dispersion in semi-major axes, produces a thin 'tube' of material spread all around the parent asteroid's orbit. Over the next few thousand years, secular precession randomises Ω, the ascending nodes of the elliptical orbits, producing a toroidal distribution, the extremes of which are observed as dust bands.

The main driver of secular precession is the gravitational effect of Jupiter, but all the planets have an effect. Over the next 10^5 years, Poynting-Robertson drag lessens the eccentricity of the orbits, circularising them, and steadily reduces their semi-major axes. Figure 1.12 shows the predictions of Grogan *et al* for the evolution in inclination and ascending node of a typical debris ring as a function of time, this time from a collision in the Koronis family. It can be seen that the end result, is a randomised I_f and Ω, such as might contribute to a main continuum, the dust simulated in "Fan" models.

The predictions of Grogan, Dermot and Durda's models favour the argument that these features, the dust bands, are *non-equilibrium* features of the cloud, i.e. they are the result of individual, random, catastrophic collision events, rather than the gradual grinding down of asteroid family members.

1.4.2.2 ISO (Infrared Space Observatory) data

In spite of its faintness in optical wavelengths, the Zodiacal Light is the dominant source of the medium-infrared sky brightness as seen from the Earth. The cross-over point is at a wavelength around 3.5 microns; at wavelengths shorter than this, the major part of the energy is scattered sunlight (with Doppler-absorption lines) – and at longer wavelengths the spectrum is dominated by thermal emission of dust grains. One of the most interesting parts of the whole spectrum is in the mid infrared (between 5 and 16 microns wavelength range), where the surface brightness increases by two orders of magnitude. In addition to the practical importance of characterising this steep rise in intensity, this is the spectral regime to look for spectral signatures of the materials constituting the small dust grains – emission lines which can be simulated in the laboratory, and then recognised in the observations of the ZL. The general shape of the mid infrared spectrum is relatively well-determined now (Abraham *et al* 1997). Very little is known, however, of the variation of spectral shape over the sky. These variations may arise from temperature differences within the Interplanetary Dust Cloud, or from differences in dust composition and size distribution in different viewing directions. Leinert *et al* (2002) analysed 29 ISO/ISOPHOT spectra distributed over the accessible sky, in order to search for variations in the spectral shape. ISO carried two separate instruments, ISOCAM (with a resolving power of 40, and wavelength range 5–16 microns), and ISOPHOT (resolving power 95, wavelength range 5.9 to 11.7 microns). At such low resolutions, Leinert *et al* found that the observed spectral range was smooth and featureless, and could be fitted by black body radiation with colour temperatures in the range of 255-300 Kelvin. The variation in temperatures was modelled by geometrical distribution of dust in the inner Solar System. From their results they concluded that the present-day interplanetary cloud is well-mixed in terms of grain composition and size distribution. Later, in 2003, Reach *et al (*2003), observed the Zodiacal Light spectrum, this time with ISOCAM, in a wide range of orientations relative to the Sun, aiming to determine whether there was a silicate feature in the ZL, and whether there were spatial variations in its spectrum. They found radiation consistent with a cloud of particles of sizes 10 to 100 microns, and they did detect a 'weak feature' which they attributed to a mixture of silicates, more dominant at low elongations from the Sun than further out (see Figure 1.13, overleaf).

Reach *et al* were able to make a comparison between our Sun's ZC infrared emission spectrum and the emission from the first exo-zodiacal cloud discovered, Beta Pictoris. The similarities and differences, along with comparisons also with spectra from comet Hale-Bopp and interstellar dust, are illustrated in their diagram, shown in Figure 1.14 (overleaf). Noting the wider 'plateau' in the emission feature in (a), and the shape of the peaks in (b) and (c), it may be reasonable to theorise that ***'b + c = a'***, and

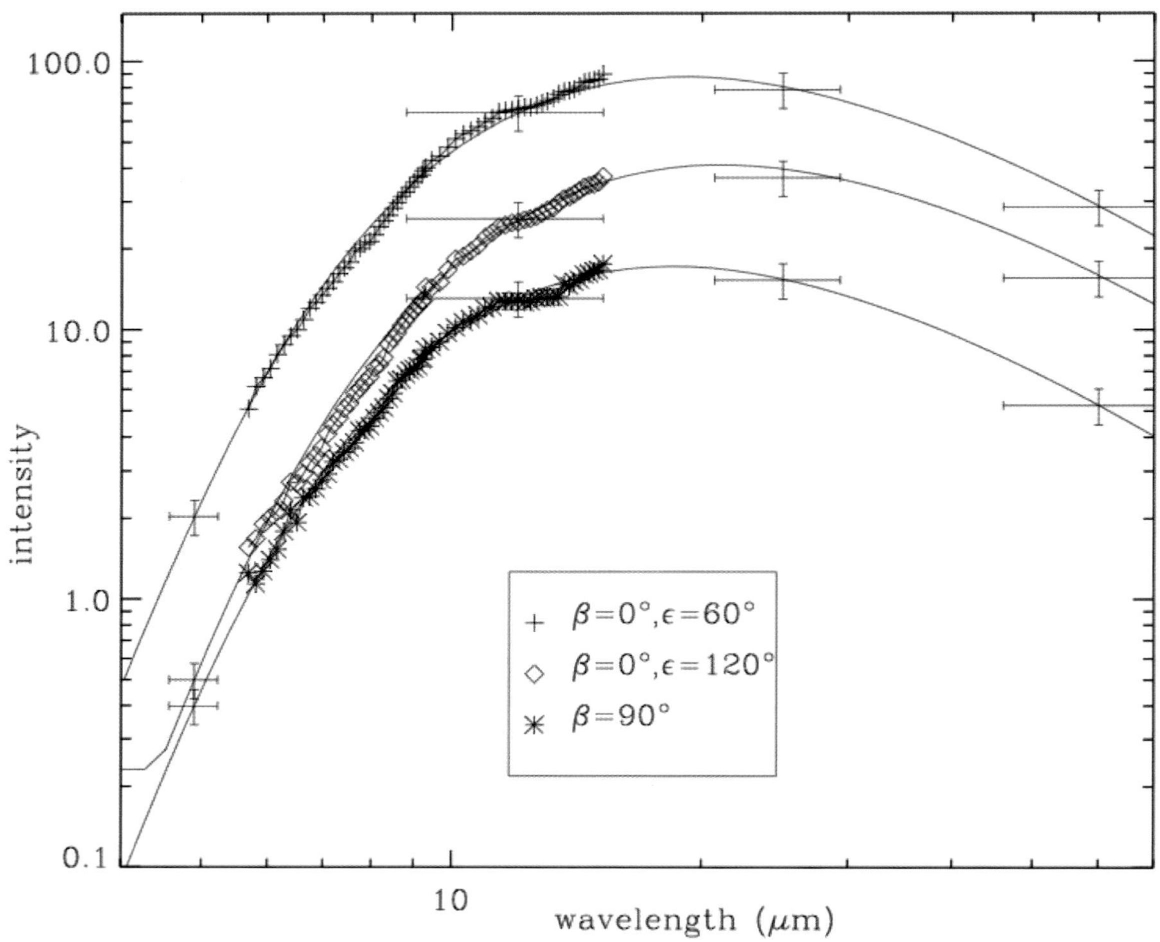

Figure 1.13 From Reach *et al* (2003). Ratio of the observed Zodiacal Light spectra to a black body continuum. The continuum was fitted using only wavelengths outside of 9-11 microns.

the feature seen here in the Zodiacal Cloud comprises elements from cometary material and interstellar material. This supports my proposition (sections 1.4.4 and 4.11) that radial velocity measurements may indicate a significant interstellar component in the dust which produces the optical Zodiacal Light.

1.4.2.3 COBE (Cosmic Background Explorer)

In 1989, the COBE satellite was launched, following the IRAS strategy, into a Sun-synchronous Earth orbit, moving in a plane which remained almost perpendicular to the direction of the Sun. It had two infrared detecting instruments on board: the DIRBE (Diffuse InfraRed Background Experiment), which had a 0.7° by 0.7° field of view pointing 30° off axis, looking at 10 infrared bands in the range 1.25 to 240 microns; and the FIRAS (Far Infrared Absolute Spectrophotometer), which had a 7° diameter field of view, pointing along the spin axis, and was sensitive to long wave infrared up to 1,000 microns. IRAS had made the discovery of the inclination of the axis of the main smooth ZC to the ecliptic, plus three sets of narrow dust bands, and a circumsolar ring at 1 AU, with a concentration of dust just behind the Earth. Kelsall *et al* (1998), using data from DIRBE, derived a detailed spatially varying Zodiacal

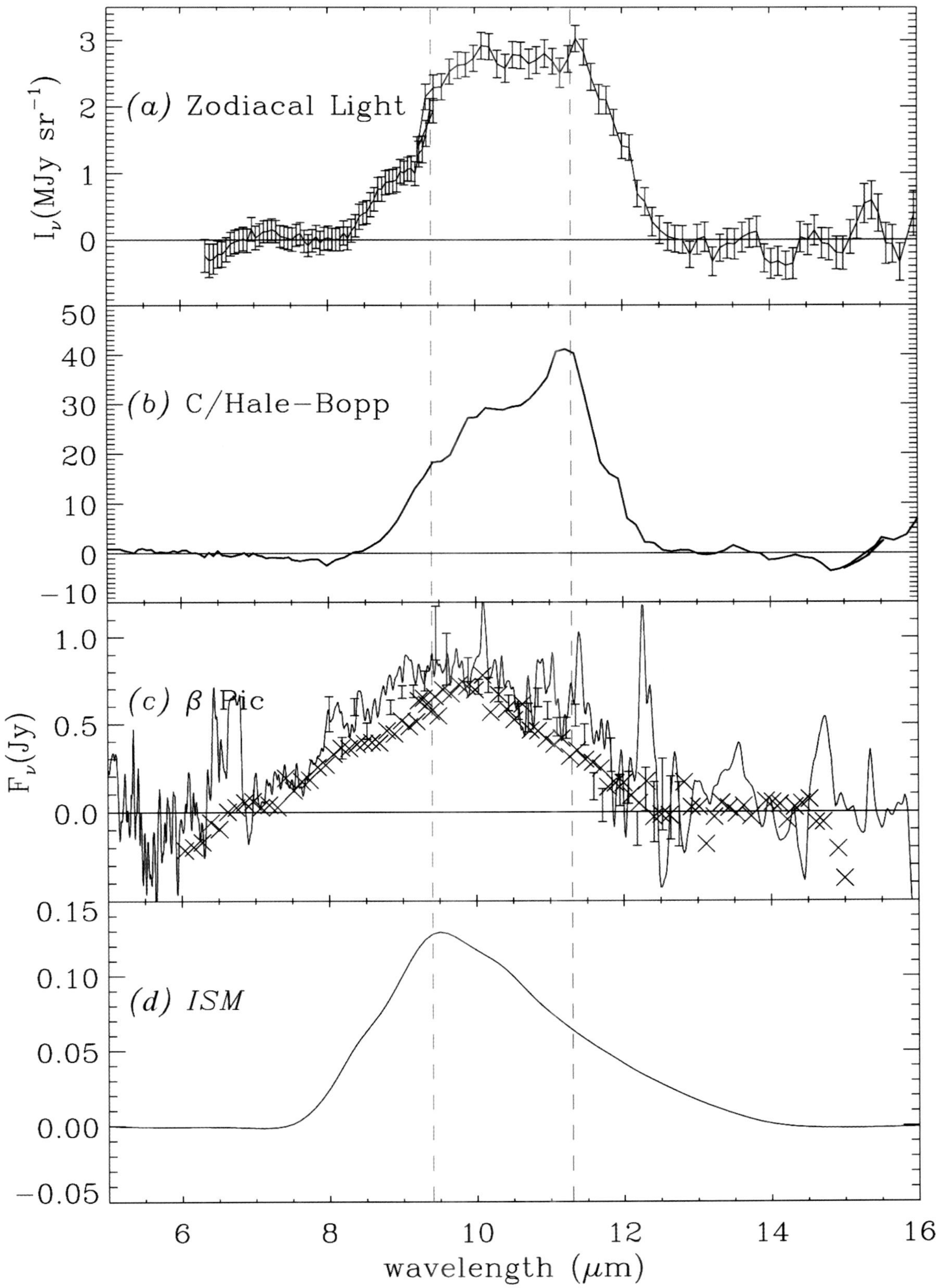

Figure 1.14 From Reach *et al* (2003). Continuum-subtracted spectra of (a) the Zodiacal Light, (b) Comet Hale-Bopp (Crovisier *et al* 1997), (c) the Beta Pictoris disc, and (d) interstellar particles (Draine and Lee 1984).

spectrum in the 1.25-240 micron range and mapped the complex geometrical structure of the ZC, the purpose being to subtract the Zodiacal contribution and accurately measure the remaining cosmic DIRB. Fixsen and Dwek (2002), combining the measurements from the two COBE instruments, built up an annually averaged infrared spectrum of the ZC spanning a huge range of wavelengths. At the lower wavelengths they found a ZL spectrum indicative of a black body at about 240° K; but at 150 microns, they found a sharp break, which they attributed to a cut-off at around 30 microns radius, which for particles predominantly composed of silicates, corresponds to a mass of 3.3×10^{-7}g. They estimated the total mass of the Zodiacal Cloud to be 2-11 times 10^{18} g, requiring a replenishment rate of 10^{14} grammes per year. The spectrum of the components of emission in the night sky which Fixsen and Dwek (2002) derive is most informative.

Figure 1.15 is adapted from their figure, which shows that around the area of the break in the Zodiacal dust spectrum, the Galactic Interstellar Medium (ISM) and the DIRB near peak intensity, but beyond about 500 microns the CMB (cosmic microwave background, echoes of the Big Bang) becomes completely dominant. This dominance continues all the way up to 5cm wavelength!

1.4.3 Collected samples and *in situ* measurements

Interplanetary dust particles of sizes 1 micron to 1 millimetre have been collected from the Stratosphere, polar ice, and from deep sea sediments (Jessberger *et al* 2001). By direct laboratory examination, it

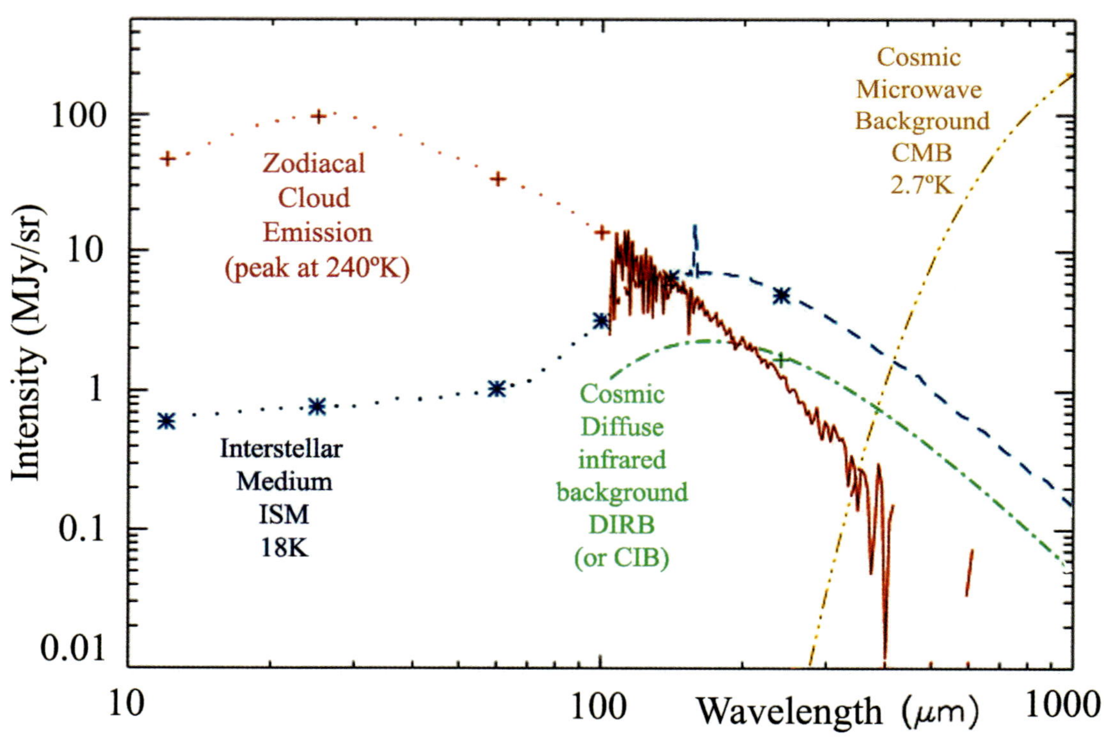

Figure 1.15 From Fixsen and Dwek (2002).
Overview of the spectrum of the night sky in the entire range of the infrared.

has been discovered that typically the particles are black, porous, and have a chondritic composition. They are complex mineral aggregates, sometimes known as 'fluffy', a mixture of large numbers of sub-micrometer sized components. One such particle, collected from the Stratosphere, is illustrated in Figure 1.16.

These individuals are believed to be reasonably representative of the population of IDP's in the vicinity of the Earth's orbit. The smallest survive unmelted because of their low entry speeds in the atmosphere, and only moderately altered by their capture and residence on Earth. One feature distinguishing these finds as a group is that they contain large abundances of noble gases, having been exposed to the solar wind and cosmic rays for a long time. Helium and Neon as well as many spallogenic isotopes are found, confirming their origin outside the Earth. Not all of the particles are fluffy – a few are smoother, but of similar composition. By measuring the abundances of two different isotopes of Oxygen in such particles, Keller *et al* (2003), ascertained that typically just one of the many tiny particles making up a fluffy IDP was of interstellar origin … the so-called Stardust. Determination of the mechanism of the capture of such a grain is a challenge! The current opinion is that the origin of IDP's has not been firmly ascertained.

Pioneers 8 through *11*, *Helios*, *Galileo*, and *Ulysses* space probes measured interplanetary dust *in situ* from 0.3 to 18 AU distance from the Sun (Grün *et al* 2001). The consensus is that within 3 AU

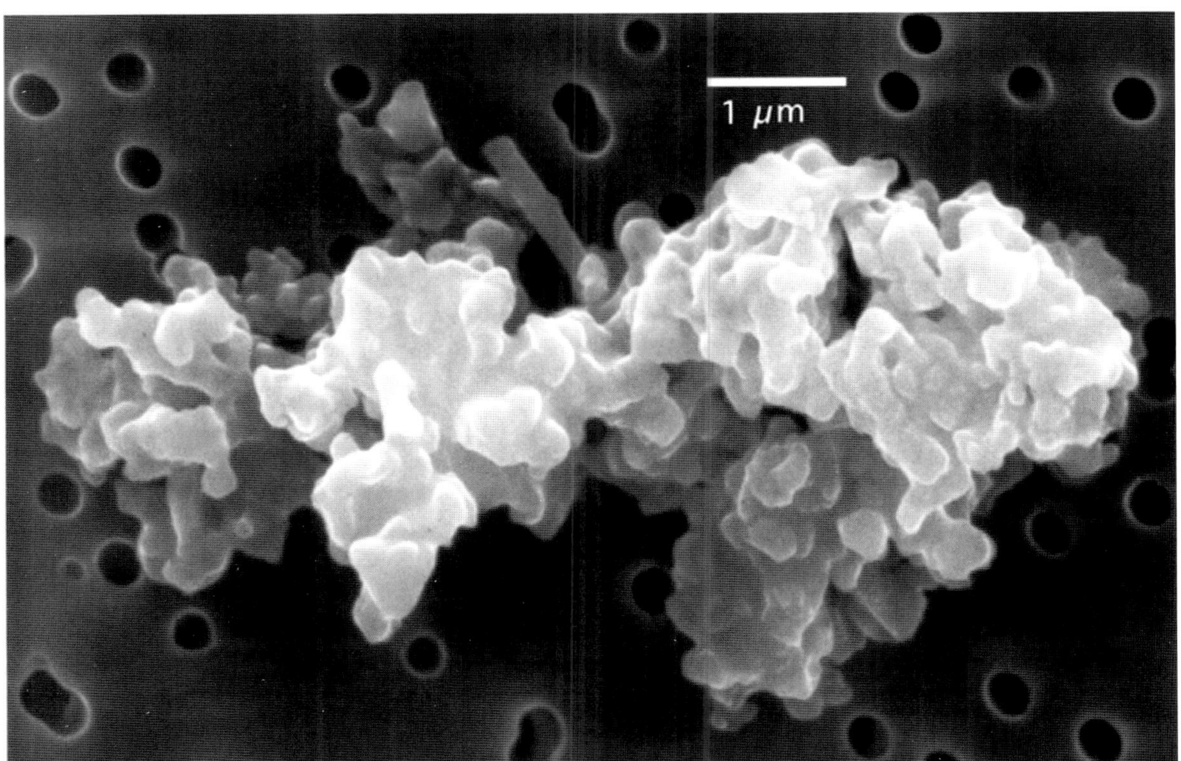

Figure 1.16 From Jessberger *et al* (2001). A porous chondritic IDP collected from the Stratosphere.

the particles orbit the Sun at low inclinations (less than 30 degrees), and their orbits are moderately eccentric (eccentricity less than 0.6). The measured dust flux displays a 100-fold decrease in density from the orbit of Mercury to that of Jupiter. In addition, particles in highly eccentric orbits have been recorded in the inner Solar System. The so-called Beta-Meteoroids are predicted to leave the Solar System on unbound hyperbolic orbits, accelerating because of the action of solar radiation pressure. Where did they come from … ?

1.4.4 Interstellar dust

One of the great 'surprises' in space observations, as noted by Sykes *et al* (2004), was that the detectors on board the space vehicle *Ulysses,* once past the orbit of Jupiter, began to register particle impacts from the opposite direction to that expected from interplanetary dust particles, and at high velocities, clearly indicating an interstellar component to the dust cloud, which predominates at these distances from the Sun, but has been now detected, by the Hiten detector, even at 1 AU (Grün *et al* 1993). If my interpretation paper had been published as planned around 1975, this discovery might have been less of a surprise! This is the source of both satisfaction and frustration to me, from the position I occupy in 2007, but a full discussion of the possibility of an interstellar component of the ZC, mirroring my writings in 1974 based on our measurements, will be offered in Chapter 4.

Grogan, Dermott and Gustafson (1996) estimated that the contribution of interstellar dust flowing into the Solar System was small – but the conclusions of Landgraf *et al* in 2000 were somewhat different. By attributing all collected particles with impact directions not attributable to elliptically orbiting dust, and speeds in excess of the local escape velocity to an interstellar component, they inferred a flux of interstellar grains with masses > $10^{-14}\,kg$ of $10^{-6}\,m^{-2}\,s^{-1}$. This is therefore an estimate of the dust density in the Local Interstellar Cloud (LIC). Confirmation of the interstellar component has also come from other quarters. Particles moving in hyperbolic orbits have been detected in 'AMOR' ground-based radar measurements by Baggaley (2000) of meteors, of limiting particle radius 20 microns. It also appears that these particles have unusual origins, to say the least – Baggaley (2000) apparently saw evidence for influx from the direction of Beta Pictoris, well-known for possessing its own 'zodiacal cloud'! Landgraf *et al* (2000), combining Radar estimates with *in-situ* measurements, found a flux for interstellar particles of mass greater than $10^{-14}\,kg$, of $10^{-6}\,m^{-2}\,s^{-1}$. I will return to this discussion in section 4.11.1.

1.4.5 Doppler measurements and models since 1974

Doppler shift radial velocity measurement was taken up again in 1978, by Fried (1978), using a different type of instrument – a correlation mask radial velocity spectrometer. This ingenious device, first devised and described by Griffin in 1967, correlates the Doppler-shifted spectrum under study with an unshifted reference spectrum, using a photographically produced optical mask, and gives a single output, recorded by a single detector, in which the information on radial velocity content is immediately

available. Fried found no difference in the magnitudes of radial velocities found in morning and evening. He also concluded from his data that the orbital velocities of the dust particles were in excess of those for elliptical Keplerian orbits, and that particles within 0.7 AU of the Sun were moving at speeds greater than the escape velocity (see section 4.13.1).

After my period of experimentation was completed, it was some years before our own Doppler-Shift based line of inquiry was continued, in a 3-year experiment, by my former supervisor Ken Reay and I. R. East, using a spectrometer of improved stability, and their further results were published in Astronomy and Astrophysics (East and Reay 1984). They confirmed the morning-evening asymmetry found in my observations, and found the material of the Gegenschein to be receding from us at 2.5 m/s. They found no evidence to support the results or conclusions of Fried: in particular they did not support the idea of hyperbolic orbits. For detailed discussion, see section 4.13.2.

After another 10-year period of little activity in the radial velocity field, the subject was revisited by Clarke *et al* (1995). They presented a more rigorous theoretical model of the ZC, and included the full quantum mechanical scattering function of the particles, as successfully used by Hong (1985) in purely photometric predictions, to predict the detailed *shape* of the absorption line. They concluded that asymmetries should be seen at high resolution, revealing the fine structure of radial velocities. In particular they offered an interpretation of the F-Corona observations of Beavers *et al* (1980), who recorded double minima in their absorption line spectra. No observations were published following on from these new predictions.

Thereafter, the line of Doppler-shift experiments seems to have lapsed, until the quest was taken up by Reynolds *et al* (2004). Their technique, using the Wisconsin Hydrogen Alpha Mapper ('WHAM') Fabry-Perot spectrometer attached to a 0.6 metre telescope, with a one-degree field of view, and a resolved width of 12 km/s (equivalent to 0.2 Å) was immediately much more powerful than our small light-gathering coelostat and filter-limited Fabry-Perot etalon. Their results have greatly reduced the error bars on the 'Radial Velocity (= wavelength shift) versus Elongation' graph, and they have been able to employ them in comparison with modern sophisticated models of the ZC with great effect. Interpretations of these new data were published by Ipatov *et al* (2006). Using measurements with rather better signal-to-noise than we were able to achieve, and higher resolution, this team have attempted to extract other parameters from the spectral curves to supplement the radial velocity estimations; they have been the first observers since my experiment to measure the behaviour of the width of the line with elongation (see Chapter 4). They also obtained spectra out of the ecliptic plane; an example is shown in Figure 1.17.

Their observations are compared with new theoretical curves calculated by Ipatov *et al* (2006). Their models compute actual spectra pertaining to a line of sight, and are based on evolutionary theories of

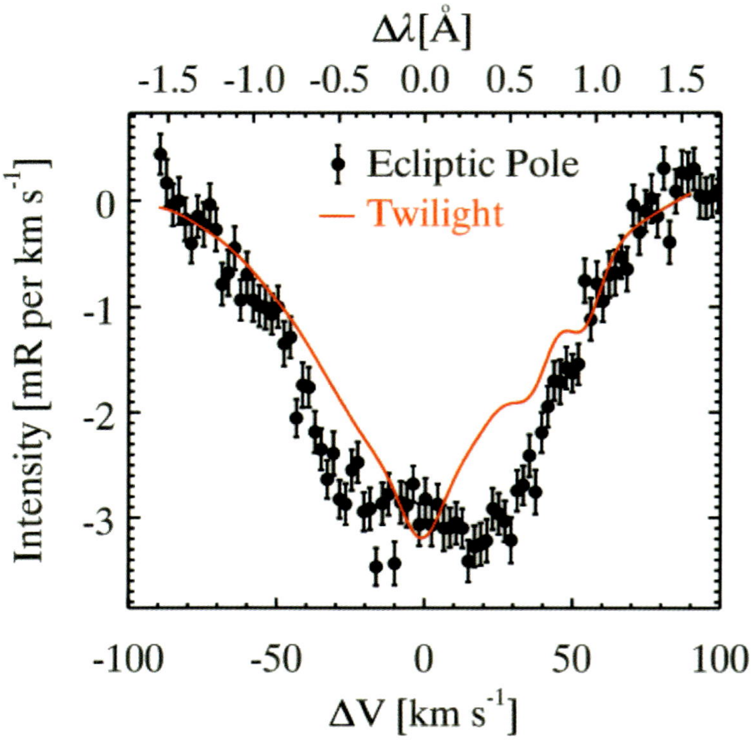

Figure 1.17 From Madsen *et al* (2005). Spectrum of the twilight sky (red) and the Zodiacal Light toward the North ecliptic pole (circles), centred near the MgI 5183.6 angstrom line.

the Zodiacal Cloud (Ipatov and Mather 2005); rather than simply starting with an arbitrary set of orbital elements for the dust particles, Ipatov *et al* go back a step to scenario in which about 15,000 asteroidal, cometary, and trans-Neptunian particles begin a journey through the Solar System. They compute the paths of these objects over time, under the influence of Poynting-Robertson drag, solar wind drag and the gravitational disruptions of the planets, and freeze the aggregate to obtain predictions of the Doppler shifted spectrum resulting from a line of sight. One of the most important parameters which controls the dynamical behaviour of dust particles is β (not to be confused with the ecliptic latitude parameter β_0). β is defined as the ratio of the Sun's radiation force to the gravitational force on the particle, and typically has values from 0.0004 to 0.4 in these Zodiacal Cloud models. It has the value $\beta = 0.573 Q_{PR} / (\rho s)$, where ρ is the density, s is the particle diameter, and Q_{PR} (not to be confused with Queen and Paul Rodgers) is the radiation pressure coefficient, which is close to unity for particles larger than one micron. For silicates of density $\rho = 2.5 g/cm^3$ the values of β 0.004, 0.05, and 0.4 correspond to particle diameters of about 120, 9.4, and 1 micron respectively. Evidently if β becomes equal to or greater than unity, the particle will be swept out of the Solar System. This is the basis for the assumption that the ZC does not contain significant numbers of particles less than 0.3 microns in diameter. Particles in exactly this balanced state of force will be able to drift through the Solar System without interference from the Sun; however they will experience perturbations from the planets! I will return to this idea in the 'continuous flow' model discussed in Chapter 4.

I will revisit the 'WHAM' recent findings in Chapter 5, and will be able to compare their findings with my measurements of 30 years earlier.

It is a remarkable circumstance for me that, after a period of relative neglect of almost 30 years, this line of inquiry, Doppler measurements of the ZL, should become active again, at the very moment when I have been able to return to the field. In the final chapter of this thesis I will outline current plans to take this study further in the near future.

Chapter 2

Preparations and experimental details 1971–1974

This chapter details the preparations for making the radial velocity measurements, and the manner in which observations were actually performed. It includes the choice of wavelength and resolving power, the site, the building of the observatory, and the construction of the Fabry-Perot Interference Spectrometer – its optics, control electronics, counting electronics, calibration equipment, and information output devices. The optimisation of the instrument is then described, and finally the techniques of operation are detailed, in the two observing periods during which my data were collected. In the following sections it should be noted that I inherited a system used in the experiments of Reay and Ring, but it soon became apparent that this was subject to inherent shortcomings and unreliabilities, and in any case needed substantial changes to operate at a new wavelength in a new, purpose-built observatory. My first year of work on the PhD was spent in redesigning and rebuilding the optics and electronics for an essentially new Fabry-Perot system, featuring a number of important improvements, which will become apparent below.

2.1 Choice of wavelength for the observations

Radial velocity measurements using the H_β region of the spectrum had proved difficult, owing to the presence of high rates of H_β emission (Hindle *et al* 1968), so I chose to study a different region: the Fraunhofer line at 5183.6 Å, present in the Sun's spectrum because of absorption by relatively cool Magnesium ions in the reversal layer. The profile of this line in pure, unshifted sunlight is shown in Figure 2.1. The hope was that this line, which is deep and relatively free from parasitic absorption by other lines in the Sun's spectrum, would NOT be present in emission in the spectrum of the ZL.

2.2 Choice of resolving power

This choice had been the result of much previous debate (Reay 1968; Smeethe 1970) which will not be repeated here. However, the outcome may be summarised by noting that if the resolving power of the

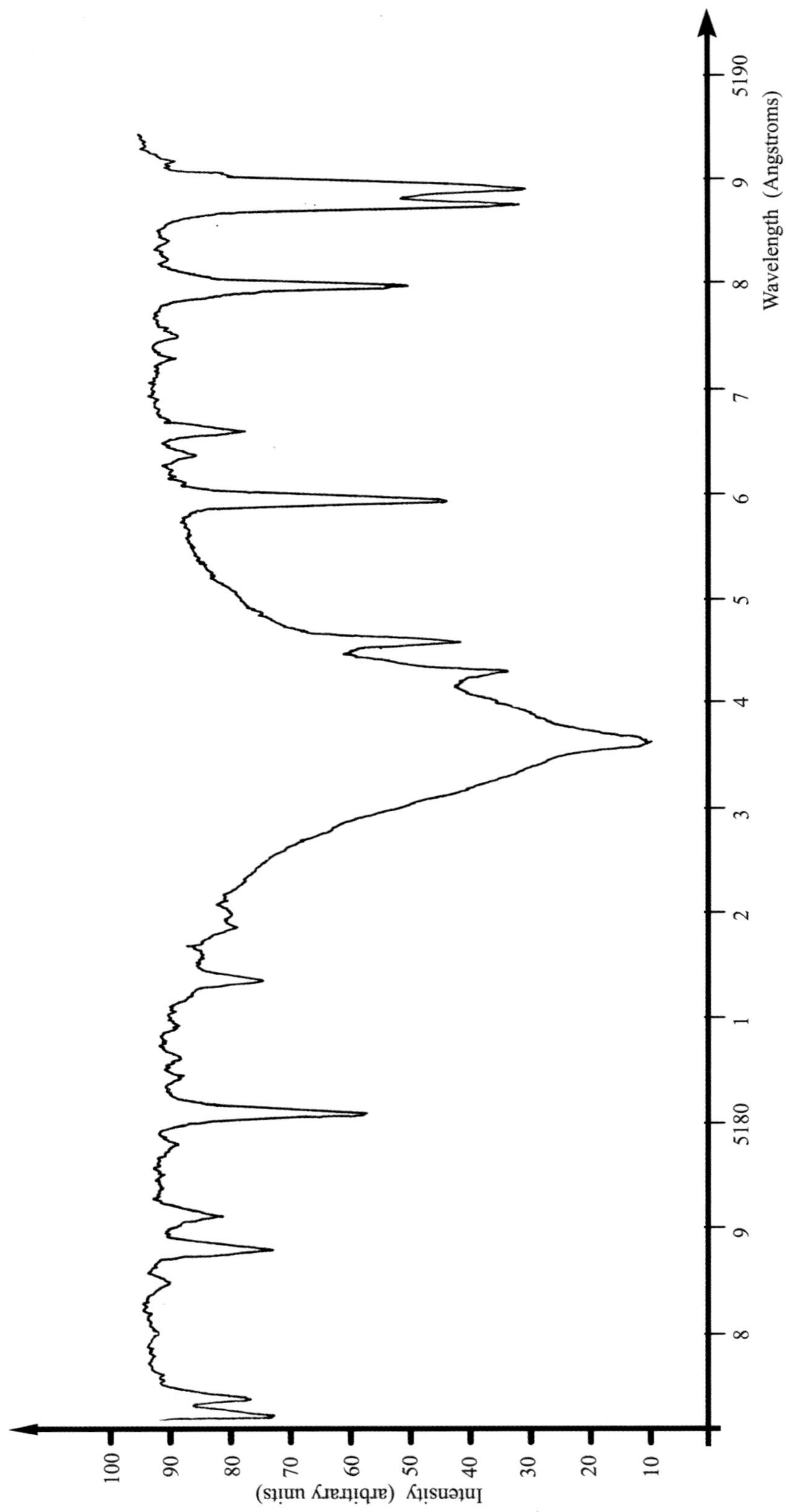

Figure 2.1 The solar spectrum in the region MgI 5183.6 Å Fraunhofer line, at about 1/50th Å resolution. (From the Minnaert Atlas, 1941.)

instrument is too low, the depth of the line is underestimated, and the exact position of the minimum point becomes inaccurate; and if the resolving power is too great, the throughput of the instrument with such a narrow profile becomes so small that very low light levels have to be measured. In this case, since only small numbers of individual photons are being counted, photon shot noise becomes dominant, and measurements of the shift in position of the minimum point of the line again become inaccurate. The optimum resolution is thus somewhere between these extremes.

A resolved width of one and a half Å was decided upon in this case, since this gave enough resolution to 'see' a well-defined absorption line, and at the same time allowed enough light through to give adequate signal-to-noise in the counting rates. It may seem strange to be making measurements on a line only about an angstrom wide with such a wide instrumental profile, but the choice for optimum resolved width has to be made based on the information sought, and the method used to elicit it. My data reduction methods are detailed in Chapter 3, and it will be seen that, processed in this way, the accuracy of the results is quite insensitive to resolving power.

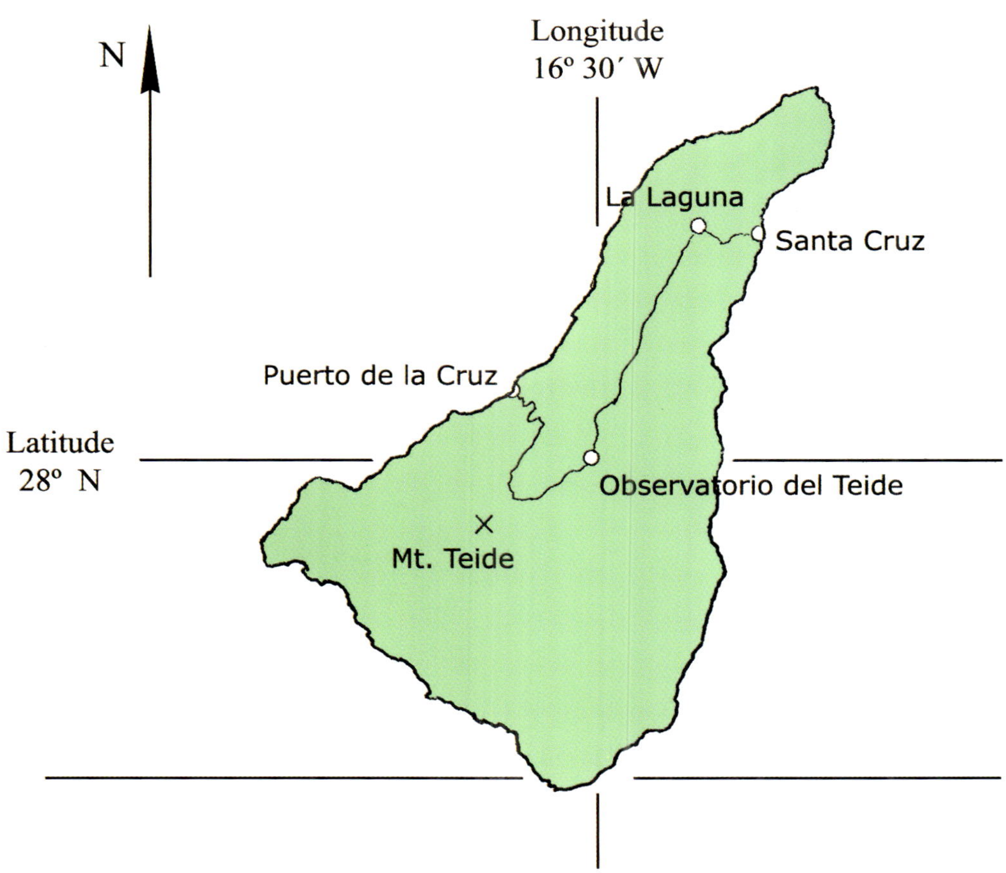

Figure 2.2 Location of the observatory on the island of Tenerife.

2.3 The site

The Testa Grigia observatory (in the perennial snow above Cervinia, Italy) having been abandoned as unusable owing to unpredictable weather, an excellent site for ZL observations was offered to us by Prof. Francisco Sanchez of the Universidad de la Laguna, at the (then) newly founded Observatorio del Teide at Izaña, situated at an altitude of 7,700 feet on a volcanic ridge in Tenerife, in the Canary Islands (Figure 2.2, previous page).

The weather conditions of Tenerife are dominated by the Trade Winds and its huge central dormant volcanic peak, known as El Teide. For much of the year, conditions are stable enough to produce an almost permanent inversion layer at a level well below the site of the observatory. The resulting view from Izaña is, more often than not, a beautiful "Sea of Clouds" below, with only occasional Cirrus clouds above. 'Seeing' is usually very good, as first ascertained in 1853 by Charles Piazzi Smythe on his famous expedition for site testing, and the only things which can interfere with good observations are dust in the air (locally known as 'Kalima'), blown in from the Sahara desert when the prevailing winds are strong enough in a particular direction, moonlight, and latterly some light pollution from the ever-growing English and German 'ghetto' of Playa de Las Americas in the South of the island, along with some new resorts being built along the South-East coast. In 1973, in our own experience, the Izaña skies were clear for about 70 per cent of the time. The site is high enough, and of suitably low latitude (28 degrees North) for the Zodiacal Light to be favourably positioned most of the time, in a good, dark sky in both morning and evening, allowing observations to be made at quite low elongations from the

Figure 2.3 The hut viewed from the South, with coelostat assembly in open position.

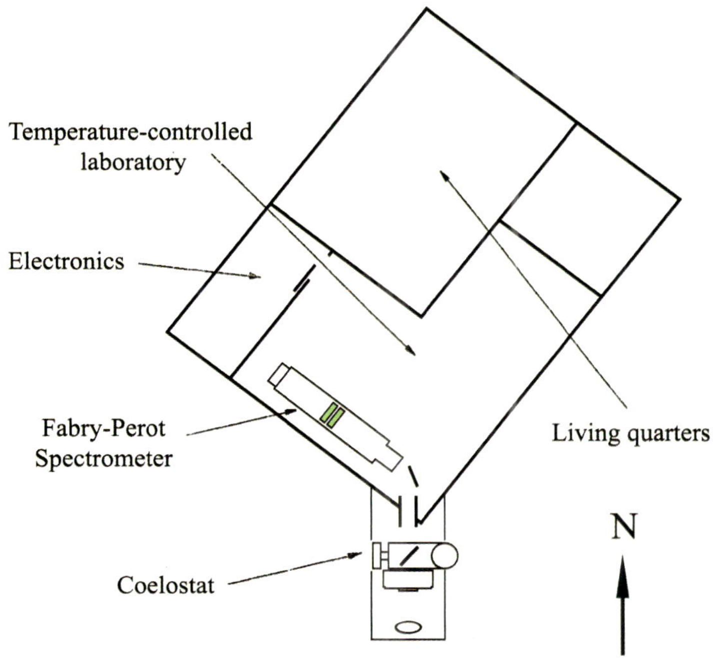

Figure 2.4 Layout of the ZL building and coelostat.

Sun. At various different times of the year, both the morning and evening Zodiacal cones stand almost perpendicular to the horizon. The observatory also had, and has, the advantage of being accessible directly by road from Santa Cruz and La Laguna, both less than an hour away. We were frequently able to observe the ZL with the naked eye, a conspicuous and splendid sight, comparable in brightness with the Milky Way. On nights of exceptional transparency the whole Zodiacal Band could be traced by eye as an extension of the morning and evening cones, a continuous faint band of light around the ecliptic, with a distinct brightening frequently discernible at the antisolar point, rendering the Gegenschein an unmistakeable though faint feature around midnight.

2.4 The ZL building

At the time of commencement of my experiment, the Observatorio del Teide consisted of just one Spanish 'hut', actually quite well appointed, built to facilitate ZL experiments mounted by the Universidad de La Laguna. It was decided to build a new hut to house the Imperial College observations. In collaboration with Professor Ring and Professor Sanchez, I designed and supervised (mainly from a distance, with the help of Carlos Sanchez on site) the building of a new 'Coseley' English hut to house our experiment (Figure 2.3). The building was modular, and specifically tailored for the ZL work, with a rudimentary living space, a thermally insulated laboratory, and a control and electronics room. The building was designed to withstand winds of up to 90 miles per hour, a necessary requirement for survival in Izaña in the Winter months. The ZL Building was situated and oriented to give optimum sky coverage to East, South, and West, with a two-mirror coelostat mounted at the South corner as shown in Figure 2.4.

Figure 2.5a The coelostat, set on concrete plinth, showing positions of its mirrors.

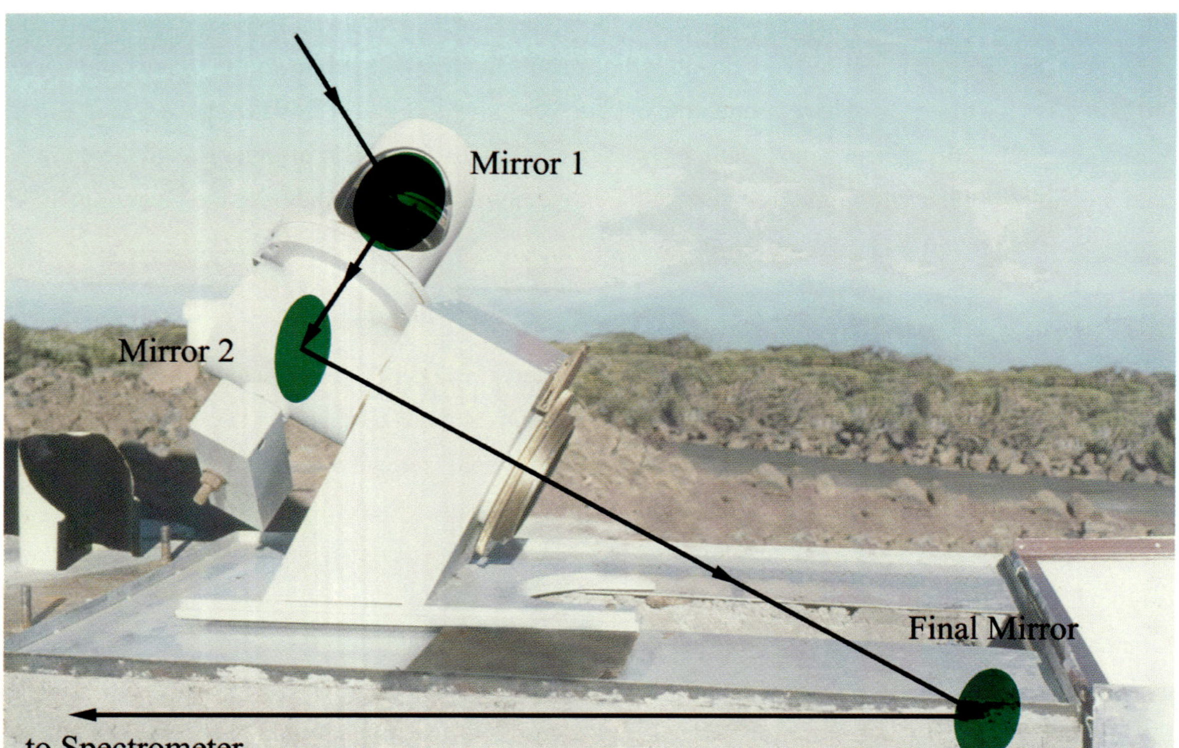

Figure 2.5b Showing the path of light through the instrument into the hut.

2.5 Instrumental details

2.5.1 The coelostat, and sky area accessible

Our interferometer was highly sensitive to movement, and it would have been impractical to attach it to a moving telescope. So instead the device of a coelostat was used, a kind of adjustable periscope containing two 6-inch aperture mirrors, to channel light from any chosen area of the sky into the Fabry-Perot instrument, which remained static inside the hut.

The coelostat mount was equatorial; Declination and RA could be manually set to areas on the celestial sphere, pre-chosen to be free from obvious contamination by bright stars, the Galactic plane or nebulosity. The field of view was approximately one degree – and the sidereal drift over the length of our scans dictated that the instrument produced spectra from areas measuring about one degree by three degrees.

The coelostat was mounted on a concrete plinth (see Figure 2.5a) with a channel in the top, forming a square section tube leading into the hut at a height of about four feet. A cover was put on the outer end of the tube, hinged so as to be able to be swung aside, providing access for lining up. In the tube was an optical bench assembly supporting a third (final) mirror to direct light from the coelostat in a horizontal beam into the laboratory, and the final f/17 lens of the coelostat, both with kinematic mounts, to facilitate lining up (Figure 2.5b). The whole coelostat was equipped with a metal weather-proof housing, attached by means of a hinge, so it could be easily swung clear for observing.

An oblique mirror in the laboratory directed the incoming beam from the coelostat on to the entrance aperture of a gas-pressure scanned Fabry-Perot interferometer. The mirror was rigidly attached via a kinematic mount to the interferometer itself.

The resulting sky area accessible to our instrument is shown in Figure 2.6a (overleaf), and shows a superb horizon from East to South, especially favourable to good observations of the morning ZL, which had been less well covered by previous experiments. Accessibility to the entire visible length of the Zodiac at any time was excellent. In the diagram the effective horizon is drawn on a projection of the sky, marked with two different co-ordinate grid systems: Altitude/Azimuth, and Hour Angle/ Declination. This diagram is a smaller version of a chart I constructed to make quick conversions between the two coordinate systems on site. Such a "Stereographic" projection of a spherical surface is constructed by placing a plane surface to touch one point of the sphere, and a point source of light at the exact opposite point on the sphere; the shadows formed on the plane surface form the projection (Figure 2.6b). This projection is convenient for accurate geometrical work because it has the property that all circles on the sphere are represented by circles on the plane surface.

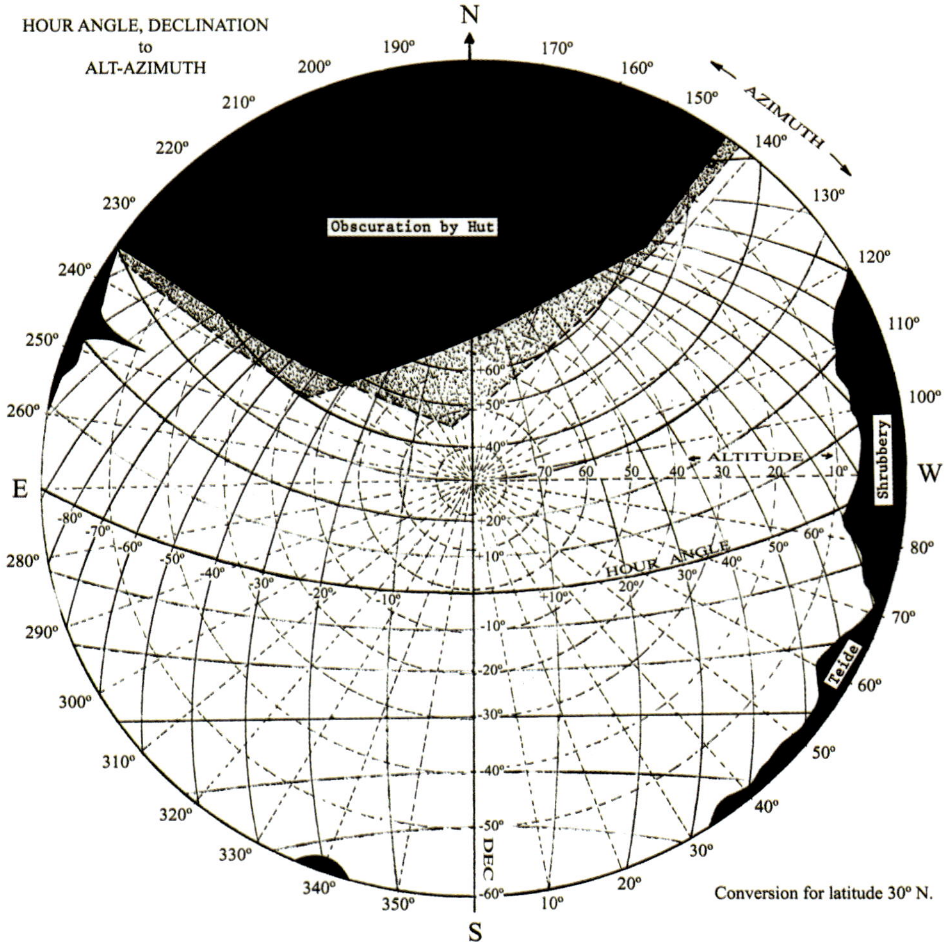

Figure 2.6a Sky accessible from coelostat marked on a coordinate conversion chart.

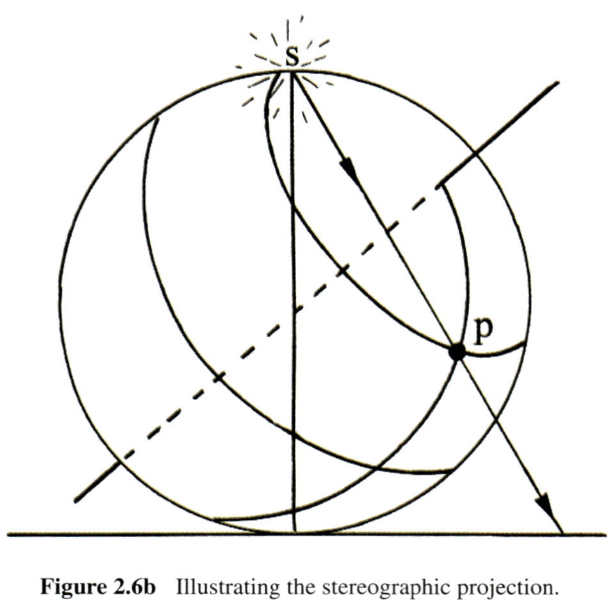

Figure 2.6b Illustrating the stereographic projection.

2.5.2 The Fabry-Perot Interferometer (F-P)

The essence of such an instrument is that light to be analysed is passed through two successive filters (see Figure 2.7). The Fabry-Perot etalon itself can be set up to allow a narrow band of wavelengths to pass through (one and a half angstroms, the chosen resolved width – see Figure 2.7a). However, as shown later, used alone, the F-P etalon would also pass a whole series of other pass-bands, separated by a wavelength interval known as the *free spectral range*. In order to suppress these unwanted wavelengths, a wider band filter (about 6 Å) is used, known as a *premonochromating filter*. The pass-band of this filter is illustrated in Figure 2.7b. The result of placing the two filters in series is shown in Figure 2.7c. As the F-P peak is scanned over the range of wavelengths to be studied, all the other peaks are filtered out by the premonochromator. The height of the peak transmission varies according to where it stands with respect to the filter shape, and this effect has to be compensated for in the reduction of the data later. The parameters shown in Figure 2.7 are roughly those chosen for this instrument. The theory of the Fabry-Perot etalon is considered in more detail later, in section 2.6.5.

A schematic of the optical system and electronics is shown in Figure 2.8 (overleaf), and the gas pressure system in Figure 2.9.

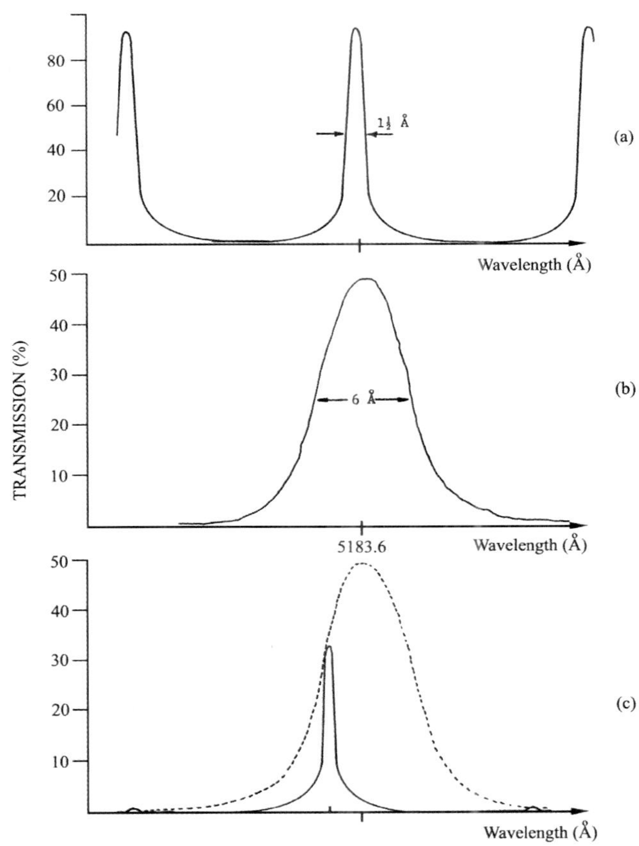

Figure 2.7 Principle of the Fabry-Perot Interferometer.

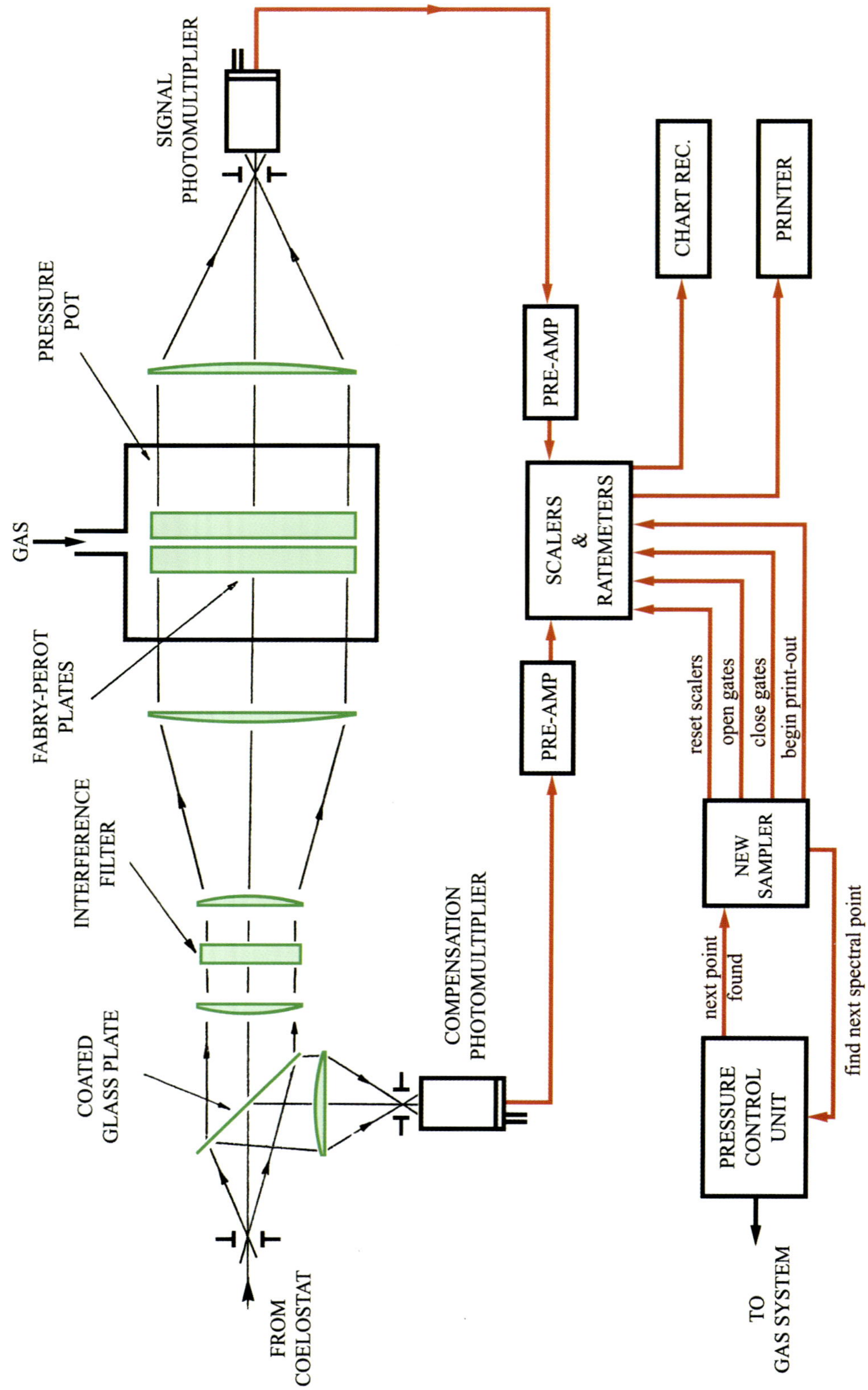

Figure 2.8 Schematic of optics, and control and counting electronics.

The Fabry-Perot etalon was housed in a pressure pot into which Nitrogen gas was fed. To scan the instrument, Nitrogen pressure was varied over a range of 6 atmospheres, allowing the instrument to scan through a range of 9 Å in wavelength. The equipment incorporated a fully automatic gas pressure stepping and cycling control, the 'pressure controller', with which the wavelength points to be sampled were selected on a programmable switchboard. Detection of the transmitted light was by a specially selected EMI 6256 photomultiplier, whose dark current was reduced to about one per second by cooling it to –30 degrees Celsius by means of a Peltier Effect cooler housing. This compares with a count rate of about 5 for the night sky. Pulses from the photomultiplier were amplified and counted by Nuclear Enterprises (Edinburgh series) scalers and ratemeters.

Compensation for sky transparency was provided by diverting a few per cent of the incident light through a gelatin filter with a 400 Å wide wavelength pass-band, centred on 5200 Å, into a second channel of pulse counting electronics. In operation, the pressure controller cycle was started by a pulse from the 'Sampler' (described below). The controller allowed the pressure to increase until a programmed point was reached, and then, after a stabilising period of 5 seconds, issued a pulse to instruct the Sampler to begin counting photons at this wavelength. The Sampler reset the scalers and allowed them to count

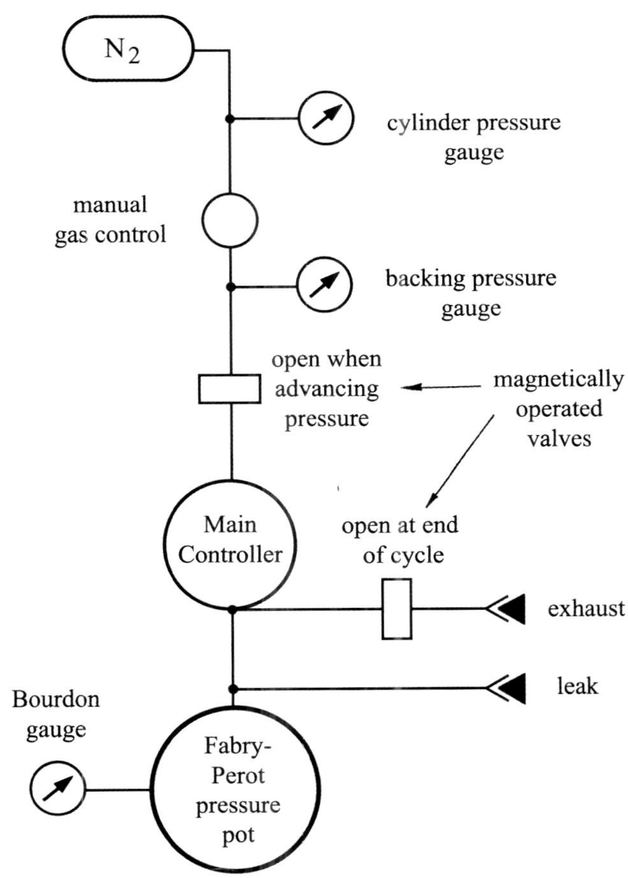

Figure 2.9 The Nitrogen system.

for a predetermined time; it then closed the gates and initiated a print-out of the total count received in each channel. When the print-out was complete, the pressure controller was automatically started off in search of the next sample point, and so on, until the end of the scan, when the controller returned the pressure pot to atmospheric pressure. Throughout the scan an analogue measure of the signal was produced by the ratemeter, which was attached to a chart reader, so a continuous visual check on signal levels was available while the scan was proceeding.

2.6 New features of the instrument

In this section I detail my own improvements, in which most of the equipment was redesigned and rebuilt.

2.6.1 Fore-optics

The front end of the F-P optics was changed so that the coelostat's 'final' lens now delivered an f/17 beam directly from the coelostat mirrors, removing the necessity for a relay lens, eliminating two glass-air surfaces and hence increasing the light throughput.

2.6.2 Field lens

A field lens was added at the entrance aperture, imaging the final coelostat lens on to the F-P plates, while the coelostat lens images an area of sky on to the entrance aperture. This was designed to prevent anomalous results due to any non-uniformities in the sky appearing on the plates, which themselves are not perfectly uniform. Also, because the plates were now at a pupil of the system, maximum throughput was achieved for a given plate area, giving increased luminosity.

2.6.3 Beam splitter for compensation channel

The glass plate which is used to deflect a small percentage of the light into the compensation channel has to sample the full area of the beam, but since this channel functions over a 400 Å wide pass-band, little intensity is needed, and in this situation it is vital to steal as *little* light as possible from the signal channel. The plate was thus coated with a single dielectric layer for maximum transmission at 5000 Å, at 45°, an unusual specification. The thickness of the layer was calculated as below. Figure 2.10 shows the geometry.

Remembering that the glass plate is to be positioned at an angle of 45° to the incoming beam, we have immediately, by the laws of refraction

$$n_g \sin\theta_g = n_l \sin\theta_l = \sin 45 = \frac{1}{\sqrt{2}}$$

where n_g and n_l are the refractive indices of the glass and the layer respectively. If the thickness of the layer is t, maximum transmission occurs when the path difference between the two reflected rays

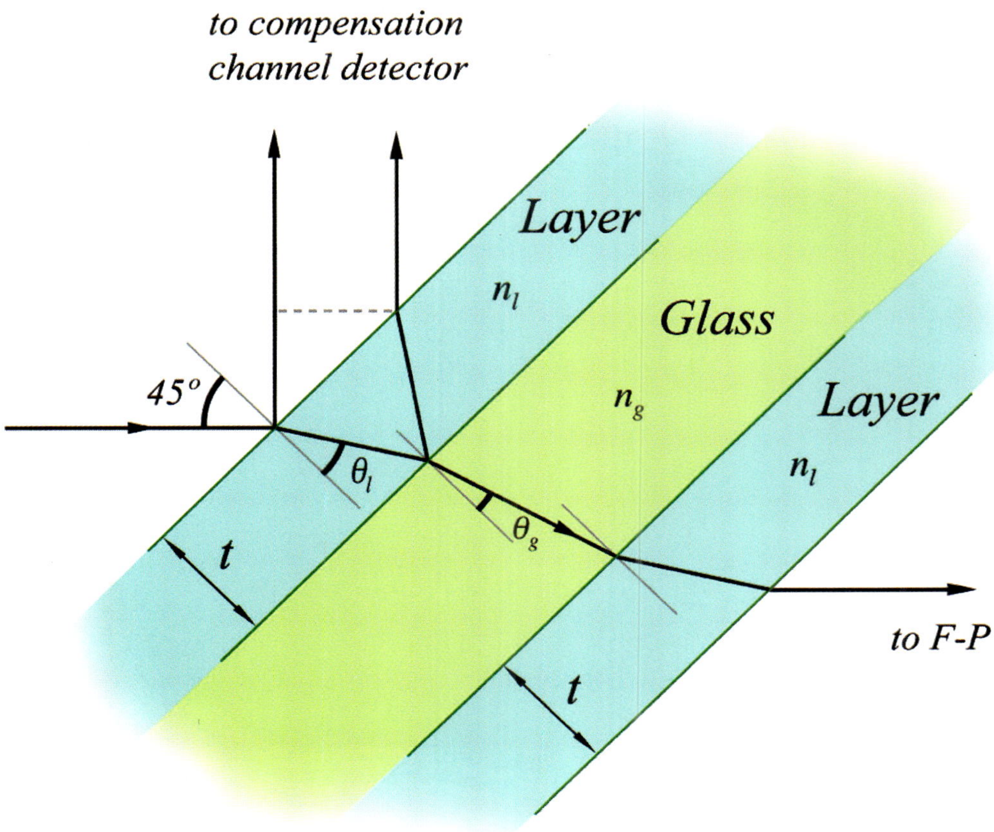

Figure 2.10 Coating for maximum transmission.

$2n_l t\cos\theta_l$ is exactly one half wave, making them interfere destructively:

$$2n_l t\cos\theta_l = \frac{\lambda}{2}$$

$$\therefore\ t = \frac{\lambda}{4}.\frac{1}{n_l\cos\theta_l}$$

$$= \frac{\lambda}{4n_l\sqrt{1-\left(\frac{1}{n_l\sqrt{2}}\right)^2}}$$

$$= \frac{\lambda}{4n_l\sqrt{1-\frac{1}{2n_l^2}}}$$

Figure 2.11 The centre part of the Fabry-Perot box, showing the etalon in its pressure pot, and remote adjusting screws behind.

The material used was Magnesium Fluoride for which $n_l = 1.37$, and for a λ of 5183.6 Å, the formula gives a thickness of $0.1\,\mu m$ (microns) for the layer. This, when applied to both surfaces, reduced the percentage of diverted light to one half of that diverted by the previous plain glass plate, further increasing the throughput of the spectrometer.

2.6.4 Construction of housing

A new light-proof cover was made for the instrument, part of which is visible in Figure 2.11, with sides which could be slid off for easy access to the optics without disturbing the alignment of the etalon. A cam mechanism was made for tilting the broad band interference filter, to enable its wavelength peak to be 'tuned' without opening the cover. Some new lateral adjustments to the lenses made it possible to line up the whole system more accurately. Most importantly, the three control screws for adjusting parallelism of the etalon were re-mounted remotely, on top of the box (Figure 2.11), where they could be operated by an observer at the 'Maxwellian' view of the plates in transmission, the eye located on the optic axis by an iris diaphragm. From such a position one can immediately see if the plates are in alignment, because the illumination is uniform, or as close to uniform as the perfection of the polishing and coatings allows. Adjusting the screws while looking at the Maxwellian view, it became easy to get a 'feel' of the etalon, and much realignment time was saved during observations. An improvement in

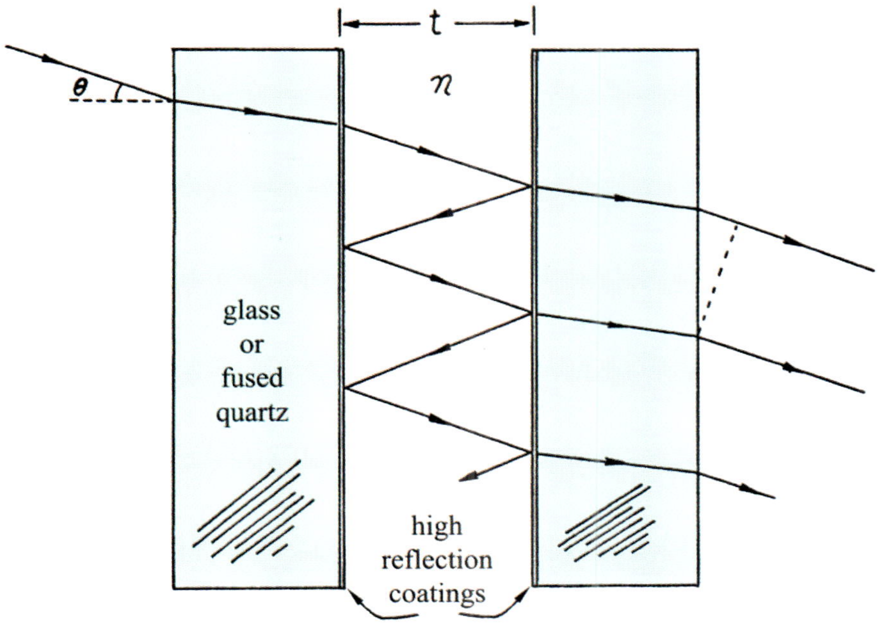

Figure 2.12a The 'perfect' Fabry-Perot etalon.

the response of the adjustments was effected by replacing the metal runners on which the plates rested by PTFE (Teflon) strips, reducing friction and lessening long period drift.

2.6.5 Fabry-Perot theory, and the finesse of a practical interferometer

The F-P plates are the heart of the spectrometer, and upon their quality depends the quality of the results. Since a new operating wavelength had been chosen, 5183.6 Å, the opportunity was taken to have some new plates made. They were cast from fused quartz, four inches in diameter and about an inch thick, polished to an accuracy of $\lambda/200$ and coated with 7-layer dielectric (alternating Magnesium Fluoride and Zinc Sulphide) stacks to give a reflectivity of 95 per cent at 5000 Å. These materials are still used today. Some recourse to Fabry-Perot theory is necessary here to explain what performance was expected, and what was obtained.

2.6.5.1 Reflection Finesse N_R

Figure 2.12a shows a 'perfect' F-P - two perfectly flat plates with a high reflection coefficient R, accurately parallel and separated by a thickness t of dielectric material of refractive index n. Any textbook of light (e.g. Jenkins and White) shows that a ray of light incident at an angle θ to the normal emerges as an infinite number of multiply- reflected rays, and the superposition of these rays, known as the *Airy Summation*, (see Appendix 7) gives a resultant transmission τ, given by

$$\tau = \frac{(1-R)^2}{1+R^2-2R\cos\Delta} \qquad \text{.........} \qquad (2.1)$$

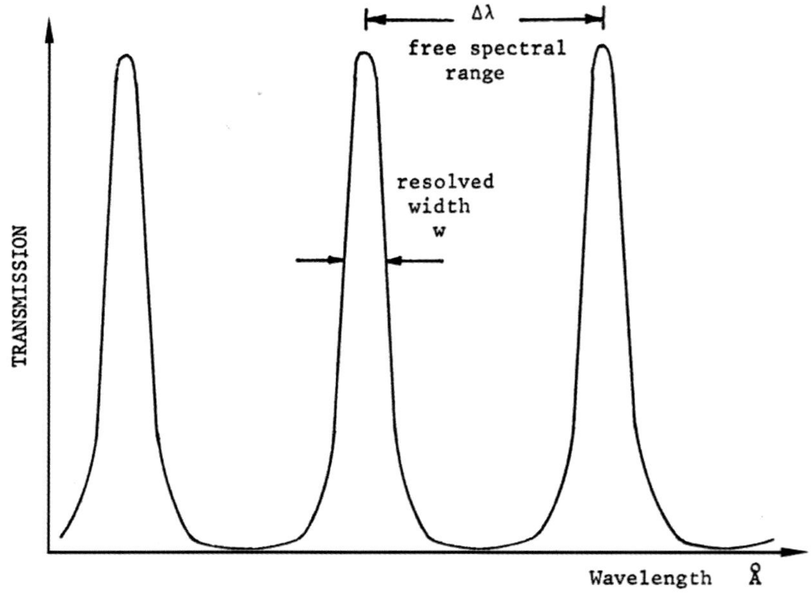

Figure 2.12b Transmission function of the perfect Fabry-Perot etalon.

where Δ is the path difference between successive transmitted rays. The geometry is the same as in Figure 2.10, and the path difference is given by

$$\Delta = 2nt\cos\theta\,.$$

Maxima of τ occur when $\dfrac{\pi\Delta}{\lambda} = M\pi$ where M is an integer, called the *order*.

i.e. $$\frac{\pi.2nt\cos\theta}{\lambda} = M\pi$$

∴ ***For a maximum,*** $$2nt\cos\theta = M\lambda_{max} \quad \ldots\ldots\ldots\ldots\ldots \quad (2.2)$$

Used as an interferometer, the Fabry-Perot is worked at around zero angle of incidence, so

Normal Incidence $$2nt = M\lambda_{max} \quad \ldots\ldots\ldots\ldots\ldots \quad (2.3)$$

A plot of τ as a function of λ has the classic Fabry-Perot shape shown in Figure 2.12b – a system of narrow peaks in transmission, separated by regions of almost no transmission at all. As was shown above (Figure 2.7), all but one of these peaks is blocked by the premonochromating filter, allowing just one narrow band of wavelengths to pass through the system.

The separation of adjacent maxima, $\lambda_1 - \lambda_2$, is called the Free Spectral Range, $\Delta\lambda$, of the etalon. For these maxima, using equation 2.2,

$$2nt = M\lambda_1$$

and $$2nt = (M+1)\lambda_2$$

so, eliminating M, $$2nt(\lambda_1 - \lambda_2) = \lambda_1\lambda_2 \approx \lambda^2$$

(since λ_1 and λ_2 are both approximately equal to the central wavelength λ).

∴ ***Free Spectral Range*** $$\Delta\lambda = \frac{\lambda^2}{2nt} \qquad \text{(2.4)}$$

A constant called the reflecting finesse N_R is defined as

$$N_R = \frac{\Delta\lambda}{w} \qquad \text{(2.5)}$$

where w (Figure 2.12b), is the full width at half power of the peak, also called the *resolved width*. N_R is a constant for a given set of plates, so this ratio is also constant.

$$N_R = \frac{\Delta\lambda}{w} = \frac{\pi}{2\sin^{-1}\left(1/\sqrt{F}\right)}, \qquad \text{where } F = \frac{4R}{(1-R)^2}$$

Reflecting Finesse $$N_R = \frac{\pi\sqrt{R}}{1-R} \qquad \text{(2.6)}$$

If the gap t is increased, or if the refractive index n of the material in the gap grows in value due to increased gas pressure, the whole pattern of peaks in transmission moves down along the wavelength scale, keeping its proportions; $\Delta\lambda$ decreases at the same rate as w because the ratio between them is a constant, the finesse of the plates. Hence for a given pair of plates we can obtain better resolution only by decreasing the free spectral range, which necessitates a narrower filter to cut out unwanted orders, and so cuts down the range of wavelengths over which the instrument can scan. Note that, in equation 2.6, R is a quantity close to unity, so $\sqrt{R}$ is even closer, and a good approximation for $\pi\sqrt{R}$ is 3. So as a rough guide,

$$N_R \approx \frac{3}{1-R} \qquad \text{.................. (2.7)}$$

Now N_R has been carefully designated the *reflecting* finesse, because in a practical Fabry-Perot there are other factors which affect the ratio between $\Delta\lambda$ and w.

2.6.5.2 Aperture finesse

The first of these is brought about because there must be a finite angle of acceptance at the plates in order to produce a finite intensity at the detector. In other words we are no longer concerned with normal incidence only. For each angle of incidence a slightly different wavelength is allowed to pass, so the result from a finite cone of light accepted at the plates is a 'smearing' of the transmission peak. The greater the angle of the cone, the wider the peak, and the more the effective finesse is degraded. An *aperture finesse* is defined as the ratio of $\Delta\lambda$ to the width $\delta\lambda_A$ of the smeared peak in the absence of any other effect (*i.e.* assuming infinite reflection finesse). If i is the angle made with the normal by the extreme ray of the cone, we have, from equation 2.2,

$$2nt\cos i = M\lambda_1 ,$$

or if i is small ...

$$2nt\left(1-\frac{1}{2}i^2\right) = M\lambda_1 .$$

but, for the central ray,

$$2nt = M\lambda_2$$

$$\therefore \delta\lambda = \lambda_2 - \lambda_1 = \frac{nti^2}{M}$$

so the aperture finesse

$$N_A = \Delta\lambda / \delta\lambda_A$$

giving, using eq. 2.4:

$$N_A = \frac{M\lambda^2}{2n^2t^2i^2} \quad \text{(In which once again } \lambda_1 \approx \lambda_2 \approx \lambda\text{)}$$

Aperture Finesse $$N_A = \frac{\lambda}{nti^2} \quad \ldots\ldots\ldots\ldots \quad (2.8)$$

Note that this has defined a throughput ($\Phi = A\omega$) of the plates where A is their area and ω is the solid angle of the cone.

So $$\Phi = \frac{\pi D^2\omega}{4}$$

where D is the diameter of the plates.

Throughput of F-P plates $$\Phi = \frac{\pi^2 D^2 i^2}{4} \quad \ldots\ldots\ldots\ldots \quad (2.9)$$

This, for maximum efficiency, should be matched to the coelostat's throughput, which can be calculated at the entrance aperture. Here, if the diameter is d and the f/ number of the beam is f,

$$\Phi = A\omega = \frac{\pi d^2}{4} \cdot \frac{\pi}{4f^2}$$

$$= \frac{\pi^2 d^2}{16f^2} .$$

For matching, the throughputs are equal, giving …

$$\frac{\pi^2 D^2 i^2}{4} = \frac{\pi^2 d^2}{16f^2}$$

$$\therefore\ i^2 = \frac{d^2}{4D^2 f^2}$$

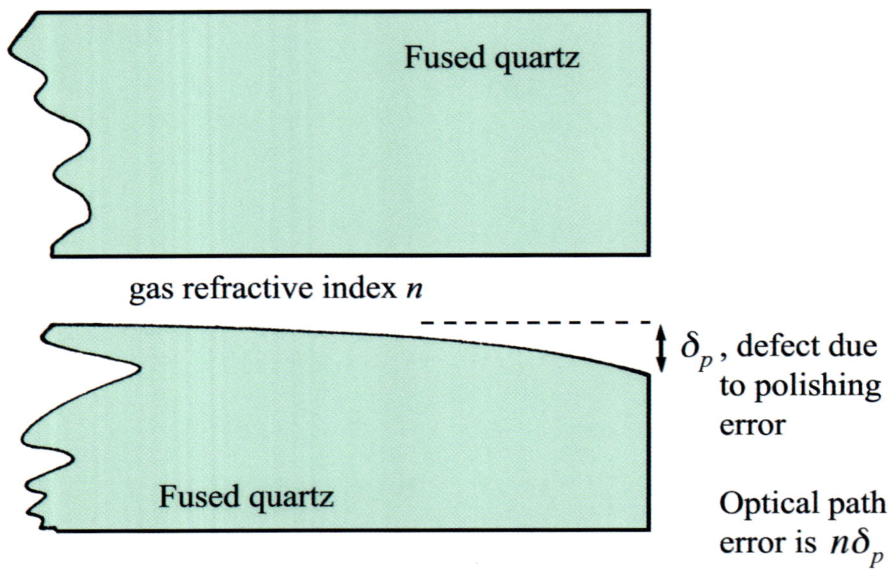

Figure 2.13a The effect of polishing errors.

Substituting this value in equation 2.8,

Aperture Finesse
$$N_A = \frac{4\lambda f^2 D^2}{ntd^2} \qquad \text{..........} \quad (2.10)$$

Thus the physical parameters of the optics alone determine the aperture finesse at which the instrument should be operated.

2.6.5.3 Defect finesse N_D

Suppose the plates are not perfectly flat but have a residual microstructure of indentations in their surfaces. They may also have a large-scale curvature or 'bowing' (Figure 2.13a). In either case a typical departure δ_p from a plane surface can be found. The usual method for determining the flatness is to misalign the plates slightly, producing the appearance of parallel bands. Measurement of the deviation of a band indicates the magnitude of a departure from flatness. This means that some parts of the plates effectively have a gap differing by δ_p from that expected, again producing a smearing of the maximum transmission peak. A defect finesse is defined in the same way as an aperture finesse, by analogy with the reflection finesse.

Defect finesse
$$N_D = \frac{\Delta\lambda}{w_D} \qquad \text{.............} \quad (2.11)$$

where w_D is the 'smeared' width due to plate defects, in the absence of any other effect. For the two extreme cases we have, from equation 2.3

$$2nt = M\lambda_1$$

and $$2n\left(n+\delta_p\right) = M\lambda_2$$

therefore $$2n\delta_p = M(\lambda_2 - \lambda_1)$$

$$= Mw_D$$

so $$w_D = \frac{2n\delta_p}{M} .$$

Hence, using equations 2.4 and 2.11, the defect finesse

$$N_D = \frac{\lambda^2 M}{2nt.2n\delta_p}$$

Defect finesse (polishing) $$N_D = \frac{\lambda}{2n\delta_p} \quad \ldots\ldots\ldots\ldots \quad (2.12)$$

Thus, given a limit to the accuracy with which the plates can be polished, there is a limit to the finesse attainable. For this reason, the defect finesse has been called by some writers a *limiting finesse*. However modern methods of polishing can give a flatness of $\lambda/200$ or better, and it is the next consideration which tends to limit the overall finesse.

2.6.5.4 Coating non-uniformities

To obtain the very high reflection coefficients needed to give high finesses, quarter-wave layers of dielectric material, of alternately high and low refractive index are deposited in a 'stack' of 7, 9, or 11 layers, etc. A seven-layer stack can give an intensity reflection coefficient of 95 per cent over a restricted spectral region (a few hundred angstroms), whereas aluminium surfaces can reach no better than 92 per cent. The layers are deposited in an evaporation chamber at a high temperature to give mechanical stability to the layer, and the plate is usually rotated and moved in an epicyclic path to produce a deposition as uniform as possible. The thickness is monitored optically, and when the centre reaches, say, a quarter-wave thickness, the plate is masked, awaiting the next layer. Any residual non-uniformity is rotationally symmetrical, and, since each process is similar, any error tends to be repeated in successive layers, giving a cumulative effect. To this may be added the effects on the substrate (glass plate) of heating and cooling, which, depending on rates of change of temperature and on how well the plates were annealed initially, can give a distortion of the substrate itself. A typical section of a non-uniformly coated plate is shown in Figure 2.13b.

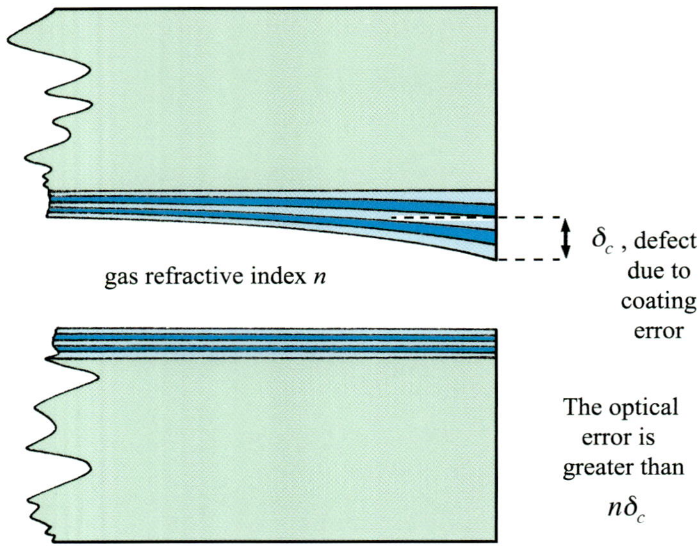

Figure 2.13b The effect of coating errors.

A fourth 'coating finesse' could be defined, but it is more convenient to add a 'coating defect' δ_c to the polishing defect δ_p and modify the defect finesse accordingly. Unfortunately the effect of non-uniformity in coatings may be more serious still, because reflections from a stack of layers differing slightly from $\lambda/4$ in thickness suffer a phase change which is also cumulative. This effect magnifies the purely geometric effect by a factor of the order of 2.

We can say that the total defect δ is given by:

$$\delta = \delta_p + \alpha\delta_c \qquad (2.13)$$

where δ_p is the polishing defect, δ_c is the coating defect, and α is the magnifying factor due to phase effects. Lissberger (1969), showed, by reducing a stack to successive single-layer problems, that in a stack of k layers, small departures δ_s in optical path from one quarter wave in the sth layer, numbered from the substrate, produce a phase change in the final reflected ray of

$$\sum_{s=2}^{k+1} L_s\delta_s$$

where L_s is the weighting factor which is greatest for layers near the gap, since the amplitude is greatest here. For most purposes δ_s is the same for each layer and so for a given system we can identify $\sum_{s=2}^{k+1} L_s$ as the magnification factor α.

The modern method of calculation is to treat the whole system using field matrices, an approach which lends itself to the use of a computer to 'solve' any configuration directly. The method gives identical results for small errors in layers.

2.6.5.5 The choice of plates

To summarise, one goes to the Fabry-Perot shop armed with three equations 2.7, 2.10, and 2.13:

Reflecting finesse $$N_R \approx \frac{3}{1-R}$$

Aperture finesse $$N_A = \frac{4\lambda f^2 D^2}{ntd^2}$$

Defect finesse $$N_D = \frac{\lambda}{2n(\delta_p + \alpha\delta_c)}$$

along with a knowledge of the resolution and scanning range required. The resulting 'real' instrumental finesse N is a complex function of all these three, given roughly by

$$\frac{1}{N^2} = \frac{1}{N_R^2} + \frac{1}{N_A^2} + \frac{1}{N_D^2}$$

N turns out to be roughly half the value of any one component finesse when all of these are roughly equal. Any mismatching brings about a waste of available resolution-luminosity product. (ref. Chabbal 1958)

For our September trip, the plates, 4 inches in diameter, were polished flat to better than $\delta_p = \lambda/100$, giving a possible polishing defect finesse of 50. The 7-layer coatings were accurate to one quarter per cent. This corresponds to a total expected error in the two stacks (14 layers in all) of one quarter per cent of $14\lambda/4$; *i.e.* about $\lambda/100$.

The magnification factor α for Zinc Cryolite is about 2. Hence

$$N_D = \frac{1}{2\left(\frac{1}{100} + \frac{1}{50}\right)} \approx 17$$

The reflection coefficient was expected to be about 94 per cent, giving a reflection finesse of 50. It was thought that the estimation of defect finesse was pessimistic and the reflection finesse optimistic, but the reverse turned out to be true. The plates suffered some deformation in coating, degrading δ_p, but the coatings gave an excellent reflection finesse, close to that theoretically possible with 7 layers.

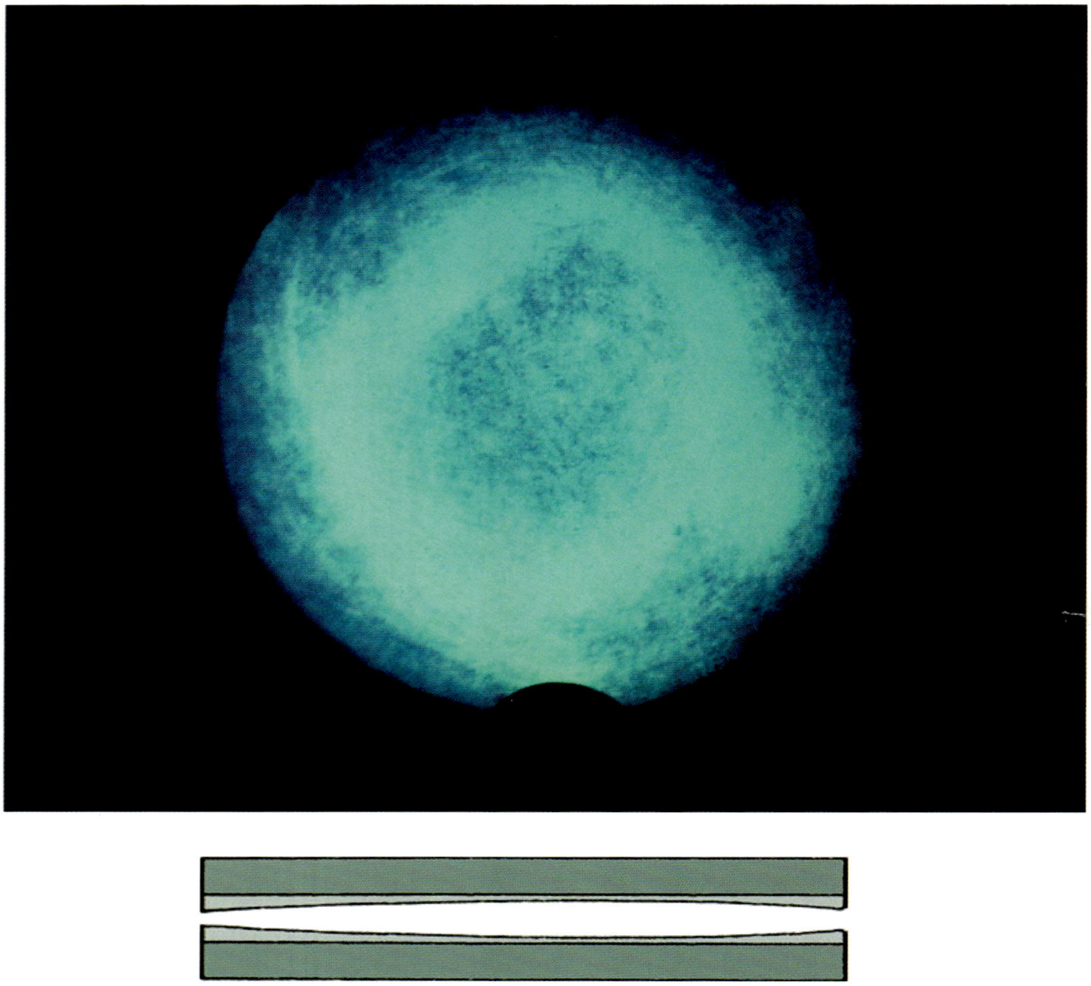

Figure 2.14 (top) Maxwellian view of F-P plates with effectively monochromatic illumination in the green, showing non-uniformity, and (bottom) possible cross-section of plates (exaggerated).

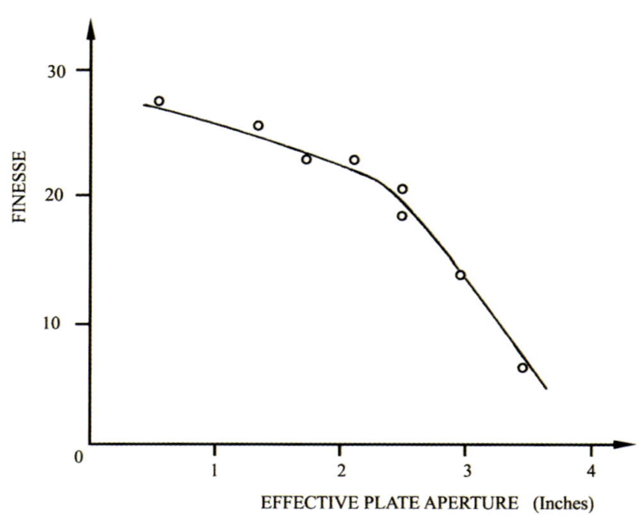

Figure 2.15 Finesses using a 2” entrance aperture (as distinct from aperture of F-P plates)

Figure 2.16 The new front end, with calibration equipment in place. The square section coelostat tube is visible at the extreme left, where light enters the hut.

Figure 2.14 shows a Maxwellian view of the plates illuminated normally with a Mercury vapour spectral lamp green line. The band-width of the illumination was much less than one angstrom. The photograph is a good illustration of a situation in which the reflection finesse is much better than the defect finesse, showing a microstructure of small scratches swamped by a long range defect indicated by a bright ring, or annulus, instead of the even illumination over the aperture that one might have expected. As the gas pressure was increased, the bright ring expanded, showing that the gap was wider at the edges than in the centre; the plates were exhibiting a convex spherical surface, rather than the perfect flat profile they were designed to have. Each small area on the plates would give a transmission peak about one angstrom wide, but the peak arrives at different pressures for different parts of the plates, so the overall peak is very wide. Masking off the edges improves the overall finesse as shown in Figure 2.15, but at the expense, of course, of throughput. A two and three quarter inch mask was considered the best compromise, and this was used in making the observations.

In this case, the excellent reflecting finesse is not only no advantage, but actually decreases the throughput.

A full discussion of the limitations in throughput of Fabry-Perot interferometers due to mismatching of finesses was given by Jacquinot (1954). Operating our plates at less than full aperture significantly decreased our counting rates, and degraded the eventual signal to noise in the scans. Thus for future similar work it may be advisable to use fewer coatings, perhaps five, in the hope of preserving the overall finesse by matching the contributory finesses more closely. The instrument was finally operated at a finesse of 10, a resolved width of one and a half angstroms, and a free spectral range of 15 Å, along with a filter width, to half power, of 6 Å.

2.6.6 Intensity calibration box

To calibrate the instrument in intensity, an attachment was made, shown in the photograph, Figure 2.16, to the front end of the spectrometer. Its construction is shown in Figure 2.17. The source was an M/F pilot bulb, calibrated at the National Physical Laboratory for luminous intensity and colour temperature at a standard voltage. The source illuminates a Magnesium Oxide screen, newly deposited, through a set of neutral density filters. The angled screen was made big enough to over-fill the acceptance cone of the instrument, reproducing the conditions of sky operation, and the intensity was arranged to be comparable with a typical signal level from the night sky, so that a valid comparison could be made. The voltage was monitored on an AVO meter, and kept constant with a rheostat. When not in use the bulb was switched off and the screen was swung out of the beam. For details of use and calculations

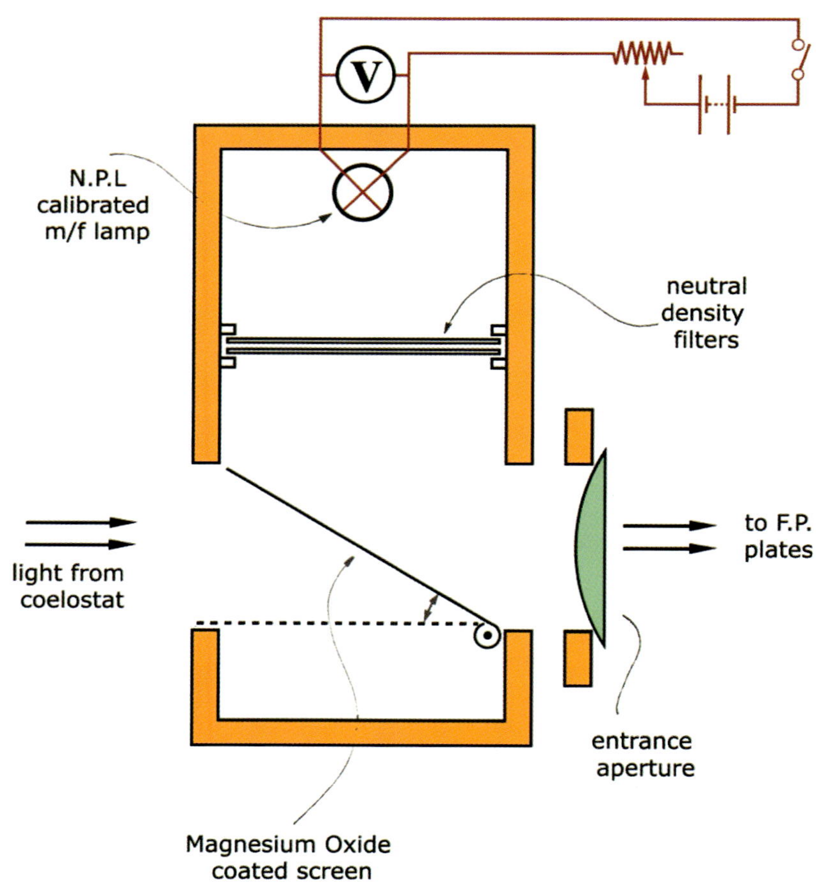

Figure 2.17 Intensity calibration box.

see Chapter 3. The intensity calibration bulb is not visible in figure 2.16; it is housed in the calibration box, as shown in Figure 2.17. The lamp visible in Figure 2.16 is a *wavelength* calibration source.

2.6.7 Wavelength calibration assembly

Two spectral lamps were used to perform calibration scans. For accurate wavelength calibration the intensity of the lamp must be constant during the time of the scan, and the illumination of the entrance aperture must be as uniform as the signal light, since evidently each annulus in the entrance aperture passes at a slightly different angle through the F-P plates, peaking at a slightly different wavelength. The former condition is not met if the lamp has just been switched on, so this lamp was left permanently switched on during a night's observing. This necessitated a rig which was easily demountable, to be hidden in a light-proof enclosure during sky scans. The assembly located a single-layer paper diffuser at the entrance aperture, illuminated evenly by the calibration lamp from a distance, via a 45° angled mirror.

The lamps used were a Phillips Zinc Halide, giving a spectral line at 5181.95 Å, and a Pye hollow-cathode lamp, giving the emission line corresponding to the Fraunhofer 5183.6 Å line, which was our designated target in the spectrum of the Zodiacal Light.

Figure 2.18 The New Sampler module, which determines integration times and generates the various pulses to operate the machine sequence.

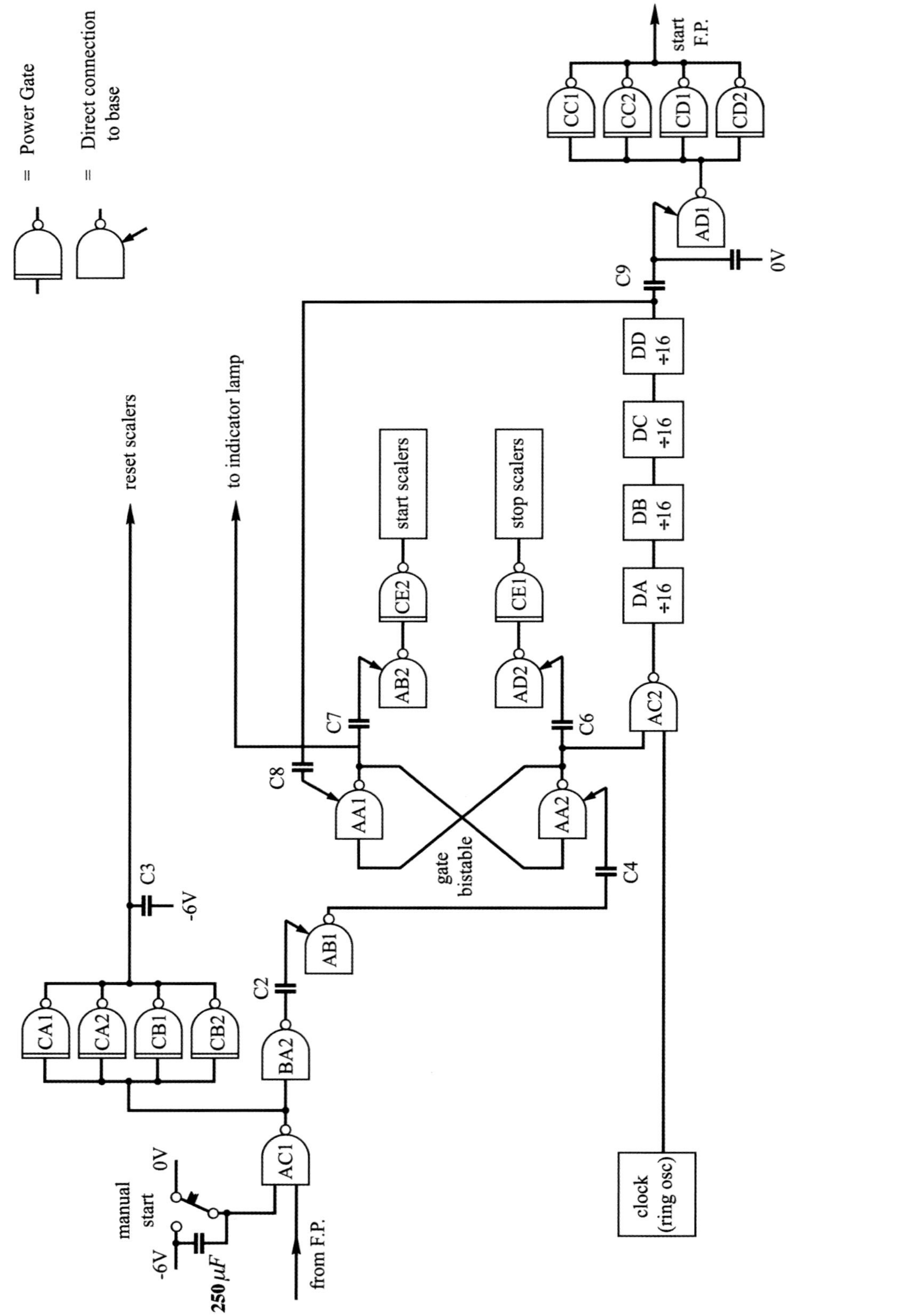

Figure 2.19 Logic schematic of the New Sampler Mk 4.

2.6.8 Control electronics

A new, reliable sampler module was constructed using MDTL logic, of my own design (Figure 2.18, previous page). The unit, whose logic diagram is shown in Figure 2.19, triggers the various functions in the control system, and incorporates a timer which determines the duration for which counts are recorded for each sample wavelength point.

The timer was completely new, designed and constructed by myself, since the old timer was the source of many problems. It is simple in design, but gave a new level of accuracy and reliability.

Instead of counting mains cycles, which can be erratic, a 'clock' waveform was generated by a ring oscillator, whose frequency was governed by a single temperature-controlled capacitor. Rather than counting pulses and decoding at various points to achieve various time intervals, the logic was set to give a signal after a constant number of pulses, and the time interval was varied by switching different controlling capacitors into the clock loop, thus changing the clock pulse rate. Intervals of one fifth, five, ten, fifty, and one hundred seconds were available. The sampler logic diagram (Figure 2.19) uses Motorola logic symbols and most of the elements perform their standard *nand* function. The short period delays, however, are a little unusual, and warrant some explanation.

Figure 2.20 shows a single element of a typical *nand* gate used as a short period delay. (a) shows the arrangement using the logic symbol for the *nand* gate, and (b) is the equivalent circuit diagram, showing the individual *npn* transistor elements inside the device. The *nand* element is simply diode logic followed by a saturating amplifier. When all inputs are floating at 8 volts (designated 'up'), both transistors conduct, and the output is 4 volts ('down'). If only one of the inputs is held 'down' at 0 volts, VT1's base becomes 0 Volts, cutting it off; VT1's output is thus down at 0 volts also, which cuts off VT2, giving a delayed output of 8 volts (up).

In the delay arrangement, a negative edge is applied via a capacitor of capacitance C directly to the base of the gate, turning the gate off and giving an 'up' output. The second plate of the capacitor now charges up through the device itself, through resistors R_1 and R_3 at a rate depending on $(R_1 + R_3)\times C$. In this case no internal leakage resistor is needed. When the second plate reaches 8 volts the gate is switched back, giving a delayed negative-going edge at the output. The technique is limited to a few milliseconds, because use of a large capacitor soon over-runs (and burns out) the previous element in the circuit; MDTL elements are essentially fast, low fan-out devices. Figure 2.20b provides a key to the internal circuitry of the *nand* gates, which, prior to the invention of multiple Integrated Circuits, were the 'state-of-the-art' elements for constructing the controllers and computers of the 1970's.

In operation, the sequence was initiated manually, or by a pulse from the pressure control unit indicating that the previous sample had been completed. The scalers were reset by a negative pulse, and after a

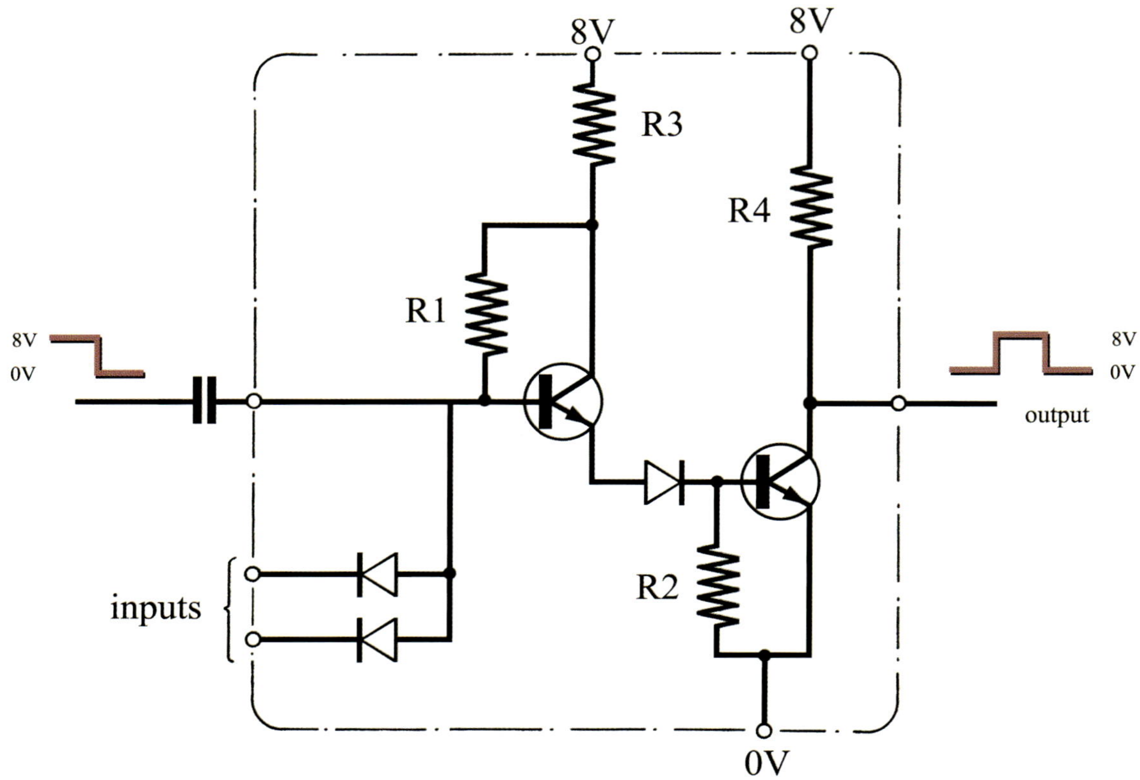

Figure 2.20a Logic representation of a short period delay.

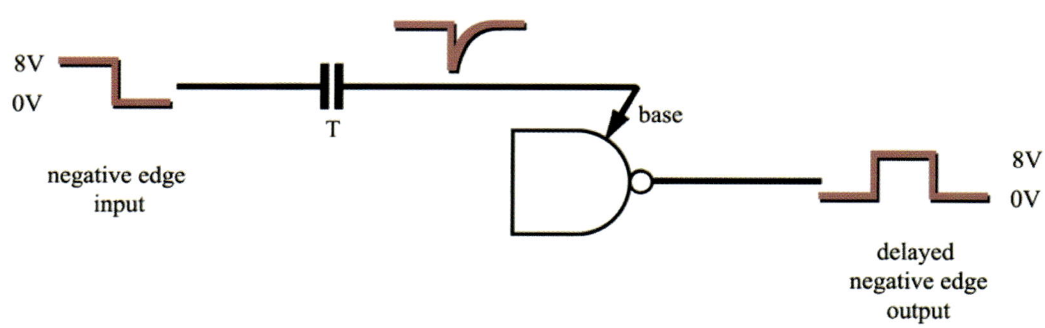

Figure 2.20b The equivalent circuit.

small delay, the 'gate' bistable was flipped to the 'open' position. This initiated a pulse to start the scalers counting signal events, and also allowed clock pulses through to the four 'sixteen' counters in series. When 16^4 clock pulses had been counted, the change in output level of the last counter was used to flip the bistable to the 'off' position, allowing no more pulses to be counted. A pulse was issued to close the gates in the signal scalers, and, after a short delay, pulses were delivered to trigger the print-out sequence and instruct the pressure controller to move to the next wavelength point. In this configuration there is fact no actual decoding of the clock pulses needed, and the counters do not have to be reset; they are already at zero, ready for the next sample.

There is an additional advantage in this system in that an error due to faulty starting or stopping of the clock can never give a large percentage error in the interval delineated. An error of a single period gives rise to a fractional error of one in 16^4. This is so small that no elaborate systems are needed to count an exact integral number of clock pulses.

The sampler was constructed in modular form so as to be compatible with the Nuclear Enterprises equipment (Edinburgh series). New power supplies were made incorporating fast voltage control cards in addition to a Claude-Lyons mechanical voltage stabiliser, and a Voltstat saturating transformer to minimise the disruptive effect of the highly erratic mains supply existing in Izaña in the early 1970's. Figure 2.21 shows how the logic elements were physically arranged within the Sampler, on four home-made printed circuit boards, plugged in to the internal frame.

The pressure controller was mounted in a new easily accessible box, separate from the programming switches and logic, minimising vibration problems during operation.

2.6.9 Output devices

To facilitate on-the-spot scrutiny of counts recorded, a somewhat unreliable paper tape punch was replaced by a Nuclear Enterprises print control unit and an Addo line printer. For each spectral point the signal count and the compensation channel count were printed on a special roll giving duplicate copies.

Figure 2.21 Showing layout of components on Sampler boards.

The analogue outputs of the rate meters were fed into a new multi-channel chart recorder along with the reference pulses generated by the pressure controller at each step, providing an excellent visual display and permanent record which could be checked for spikes and other anomalies.

2.7 Optimisation of the system

2.7.1 Optics

Much care was taken to ensure that the optics were correctly aligned. The optical bench in the channel under the coelostat was used to define a principal optic axis. The coelostat circles were set to the Sun's position accurately calculated from ephemeris tables, and the coelostat frame was positioned to project the Sun's image on to the centre of the third mirror, which was in turn adjusted to send the image exactly along the optic axis of the tube. The oblique mirror in the hut was then adjusted to project the image centrally on to the entrance aperture of the Fabry-Perot etalon. The F-P optics were already aligned by autocollimation, with the plates set perpendicular to the optic axis by illuminating them with a bright extended source and lining up the centre of the ring pattern on the exit aperture. It then only remained to make sure that the F-P axis coincided with the axis of the incoming beam, which was centrally illuminated by the Sun's image. Using this procedure it was possible to point the coelostat to an accuracy of half a degree, more than adequate for this work.

2.7.2 Counting electronics

The pulse height spectrum of a real signal differs from that arising from noise inside the photomultiplier tube, and an optimum signal to noise figure can be obtained by correct positioning of the threshold voltage level above which a signal pulse is counted, and correct setting of the EHT voltage used for the photomultiplier. The quantity to be optimised, as related by Hindle and Ibbett (1966), is

$$\frac{S}{\sqrt{S+2D}}$$

where S is the signal count and D is the 'dark current' count. To optimise this, readings of counting rates with the shutter open and closed were recorded for a number of settings of EHT and threshold levels, and the value of this formula calculated in each case, taking care to allow time for the tube to settle down after any change in EHT.

2.7.3 The system as a whole

To test the equipment as a system, numerous scans of the spectral lamps were performed, revealing the shape of the instrumental function, and how it changed in various conditions of use. Experience was gained in this way of the system's stability while undergoing vibration and temperature changes, and its response to adjustments. Drift was at its most troublesome in the early evening when the external temperature was dropping quickly. In the first observing trip this effect was minimised by aligning the etalon at sunset to give a Magnesium I emission line peak at a slightly higher position in the pressure range than that eventually required, and gradually tightening the adjustment screws to bring the peak down in pressure to the working value at the time of the first sky scan of the night. In this way the wavelength drift due to compression in the spacers and springs was crudely balanced against the drift due to temperature change, and stability was easier to obtain. In the second trip using better temperature control equipment (fan heaters with thermostats, fitted inside an enclosure made of polythene sheeting) the drift was largely eliminated. As well as spectral lamp scans and scans of the filter using a continuous spectrum white light source, scans were recorded of the day sky. These furnished a useful standard absorption line shape, which was compared with a theoretical convolved shape (see Chapter 4) to check that the correct resolution was being achieved. Test scans were also recorded of moonlight, bright stars, the Milky Way, and 'dark' sky, to check for reasonable counting rates to be expected. Attention was then turned to the Zodiacal Light.

2.8 How the observations were made

Zodiacal Light measurements can only be made in the absence of twilight and moonlight. A typical observing period begins about three days after Full Moon, when only about an hour of complete darkness is available between the end of astronomical twilight and moonrise. The available observation time then increases each night, up to a maximum at astronomical New Moon, and thereafter diminishes as the crescent 'New Moon' of every-day parlance sets later each evening. The observing period ends about three days before the next Full Moon. I have represented a number of such cycles diagrammatically in Figure 2.22, which is a useful quick guide to viable ZL observing time available in 1971. Of course clear weather is also necessary, and an absence of light pollution. Around New Moon both morning and evening Zodiacal Light can be seen. Also shown on the chart are the angular elevations of the principal axes of the morning and evening cones; the October period is seen to be most suitable for observations of the morning ZL at its maximum elevation (about 87 degrees at Izaña). For similar reasons April is most suitable for observing the evening ZL. These periods fortunately also give freedom from contamination by the Milky Way at low elongations from the Sun. I planned my observations to be mainly concentrated in the ecliptic plane, and before an observing trip a number of points along the line of the ecliptic were tabulated.

The method adopted was to choose points separated by about five degrees along the ecliptic characterised by a number according to their ecliptic longitude, and note their coordinates in Right Ascension and Declination. Points were chosen as far as possible free of nebulosity and bright stars, but were later to be improved by experiment.

The procedure for making an observation was as follows: one of the tabulated points was chosen, its suitability verified by eye and reference to Figure 2.6, and the coelostat was set to accept light from this point. Counting rates were then sampled for between 10 and 18 spectral elements over a range of about 4 Å centred on the 5183.6 Å line. The integration time for each sample point was 50 seconds. On each

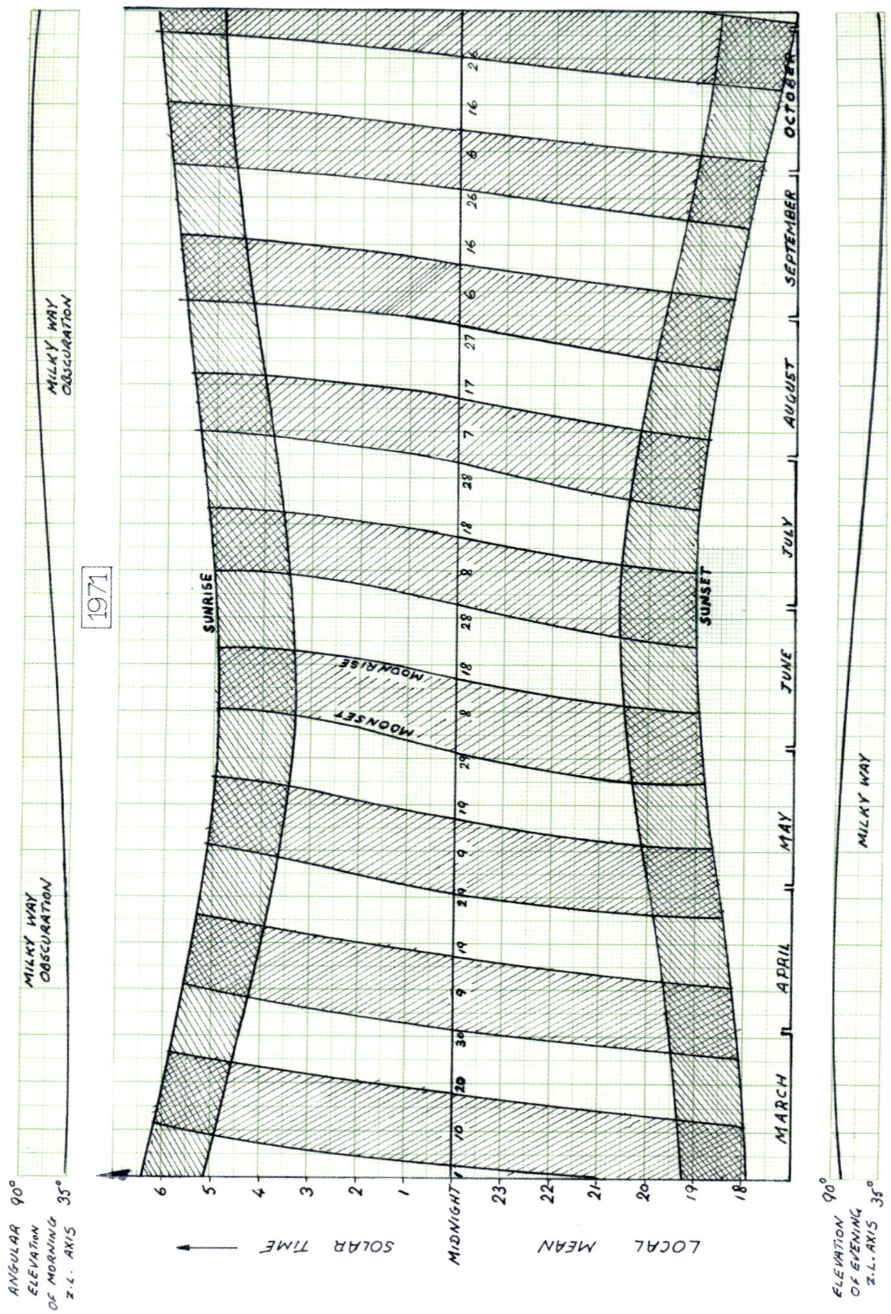

Figure 2.22 Schematic of suitable times to observe the Zodiacal Light in 1971. Each day is represented by one vertical line, and unshaded areas represent a possible dark sky. Add 1 to date after midnight

night the points available were systematically covered, using the periods close to twilight to look at low elongations. From night to night, the tabulated ecliptic points gradually shifted in elongation from the Sun, as the Sun's position changed relative to the background stars, and so over a few days a fairly uniform sampling of elongations was attained, with a deliberate concentration on the Gegenschein.

If any tabulated point showed a consistently erratic pulse count in the compensation channel, a better point was found and henceforth used instead, so by experience a good set of points was built up.

The area of the sky projected on to the entrance aperture was about one square degree, and since the duration of a scan was typically 12 minutes, the spatially resolved patch of sky was an oval three degrees long and one degree wide.

Unless special care is taken, any drift in the Fabry-Perot spacing will result in a drift in the position of maximum transmission during the scan, adding uncertainties to the wavelength calibration. The pressure corresponding to the position of the unshifted 5183.6 Å line was determined immediately before and after each scan, by scanning the Magnesium hollow cathode lamp. These are the sky scan's associated *finesse scans*, so-called because they are also used to test whether the finesse was good enough, or not, in which case the plates needed realignment. Care was taken to ensure that, in these calibration scans, the entrance aperture was evenly illuminated and that the intensity was constant, for reasons given above. Scans of a continuous white light source were made before and after each night's observations to determine the position of the filter peak in wavelength, and at these times dark counting scans were also recorded, these being performed by covering the entrance aperture rather than closing the shutters, in order to count stray light along with the photomultiplier dark or noise current. In the April 1972 observing session, the stabilities of both etalon and filter were greatly improved by temperature-controlling the instrument to $\pm 0.5^{\circ}$Celsius. Two methods presented themselves for improving the stability still further in subsequent experiments; an optically contacted etalon with quartz spacers, or variable thickness spacers controlled by a good servo-loop. Both systems were later realised in the Imperial College Infrared Astronomy Department. The servo-controlled piezo Fabry-Perot, developed by Tom Hicks, proved to be particularly rewarding in performance: in the hands of Hicks and Reay, it gave birth to a highly successful manufacturing company devoted to the commercial production of these devices for use in research and industry, and in 1984, one of these instruments enabled East and Reay to revisit ZL Doppler shift measurements (see later, section 4.13.2), and amass data with a significantly improved signal-to-noise ratio.

The wavelength scale was established experimentally by scanning the Magnesium and Zinc lamps alternately, and in Chapter 3 the use of these scans to make this calibration is described. Intensity calibration scans were performed using the attachment described above, and scanning exactly as for a sky scan. During the day some blue sky scans were recorded for comparison.

In all, 245 scans of the Zodiacal Light and Gegenschein were obtained, plus about 20 various others, including the apex and antapex of the Sun's motion in the local galaxy.

This completes the account of experimental details.

Chapter 3
Reduction of the data

This chapter contains an account of the processes applied to the raw scans, after the observations were completed, to transform them into a set of corrected spectra, suitable for the extraction of parameters to compare with theory, together with some preliminary observations on the data. I describe here the absolute calibration of intensity, and the manner in which the scale and datum position for wavelengths were determined, and how the scans were corrected for various imperfections, interferences and systematic errors. The unexpected discovery of Magnesium I emission is discussed, along with its effect on the interpretation of the data for radial velocity determinations. A system of averaging the fully corrected absorption line scans, grouped according to position in the ecliptic, is described, together with various attempts to fit curves to the experimental points. The method of extracting parameters from the finally chosen Gaussian fits is detailed.

3.1 Intensity calibration

3.1.1 Method and calculations

The source used for intensity calibration is described in Chapter 2. The energy input to the interferometer is calculated as follows.

The National Physical Laboratory supplies tables (Jones 1970) of $Q(\lambda)$, the radiant spectral intensity of a tungsten source at colour temperature T_c°K inside a glass envelope. The value I have used in the calculations is a compromise between this figure and that appropriate to a black body, since for a coiled element the inside parts behave like a black body.

The energy per unit wavelength interval radiated by the lamp is given in the tables as:

$Q(\lambda)$ microwatts per steradian per nanometre.

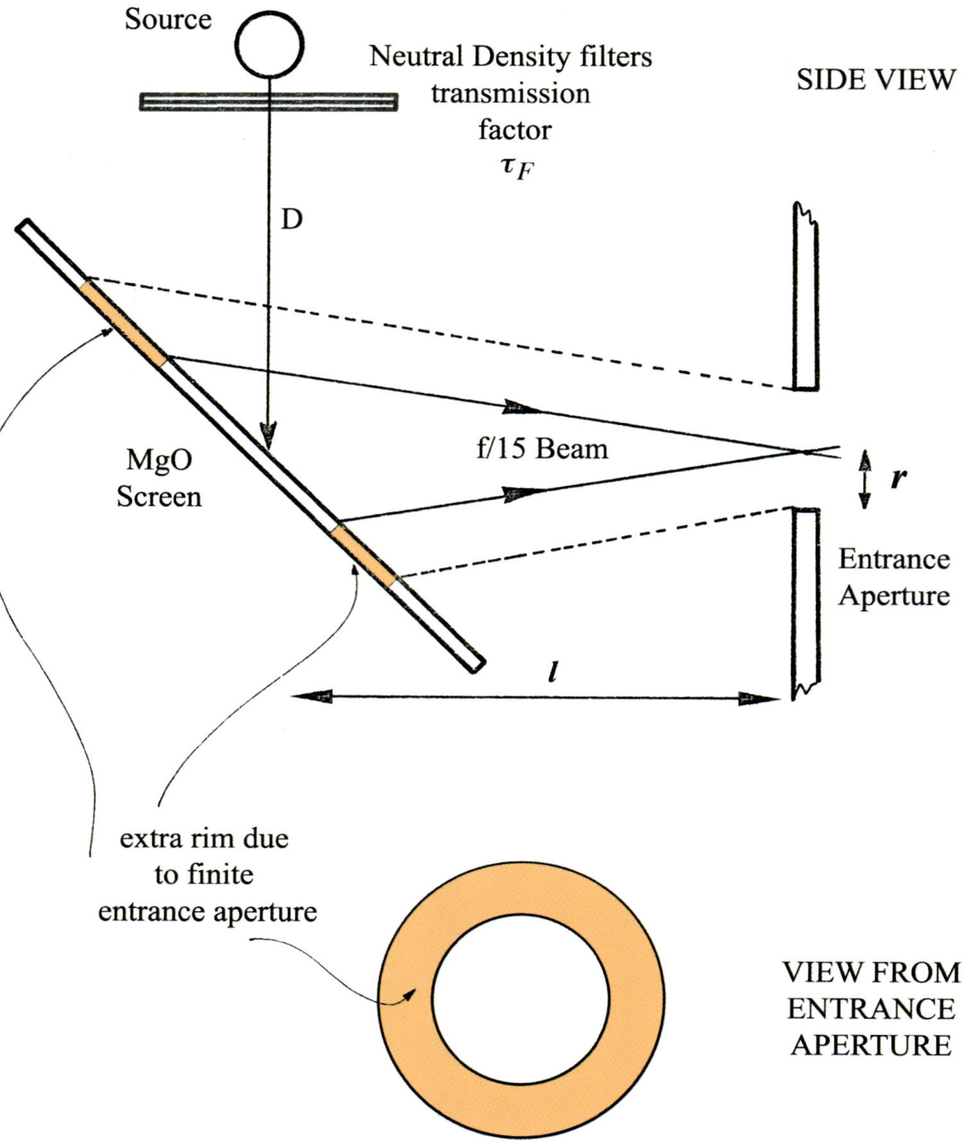

Figure 3.1 Photometry of intensity calibration.

Hence the spectral irradiance on the MgO (Magnesium Oxide) screen, placed at an angle of 45 degrees, at a distance of D cm, modified by the transmission factor τ_F of the filters is (referring to Figure 3.1)

$$Q(\lambda)\times\frac{1}{\sqrt{2}D^2}\times\tau_F\ \mu\text{watts cm}^{-2}\text{nm}^{-1}$$

The entrance aperture accepts an f/15 beam, so a small aperture would 'see' on the diffusing plate, an elliptical shape, of area

$$\frac{\pi}{4}\left(\frac{l}{15}\right)^2 \sqrt{2}\ \text{cm}^2$$

where l is the distance of the screen from the entrance aperture. However the area actually seen is much larger than this because the entrance aperture has a finite radius r cm, so the projected shape has an extra rim of width r cm (Figure 3.1). The total area seen is

$$\frac{\pi}{4}\left(\frac{l}{15}+2r\right)^2 \times \sqrt{2}\ \text{cm}^2$$

Emission from such a screen (assumed non-absorbing) at an angle θ to the normal has an intensity per solid angle $E\omega\cos\theta/2\pi$, where E is the total incident intensity, and ω is the solid angle, in this case, that subtended by the entrance aperture at the screen, which is $4r^2/l^2$. Thus the total flux received by the F-P is

$$\frac{Q(\lambda)}{\sqrt{2}D^2} \times \tau_F \times \frac{\pi}{4}\left(\frac{l}{15}+2r\right)^2 \sqrt{2} \times \frac{4r^2}{2\pi l^2\sqrt{2}} \qquad \mu\ \text{watts nm}^{-1}$$

$$= \frac{Q(\lambda)\tau_F r^2}{4\pi D^2 l^2\sqrt{2}}\left(\frac{l}{15}+2r\right)^2 \quad \mu\ \text{watts nm}^{-1}.$$

For a given finesse, a given photon counting rate represents a definite flux per unit wavelength range. A scan of the calibration source gives a measured N_{cal} counts per second at the filter. If the filter shape is normalised to unity maximum for the purpose of dividing it out from the scans, for any filter–corrected count per second N measured from the sky, the corresponding energy per unit wavelength range is (now more conveniently expressed per Angstrom):

$$W = \frac{N}{N_{cal}} \frac{Q(\lambda)r^2}{4\pi D^2 l^2\sqrt{2}}\left(\frac{l}{15}+2r\right)^2 \times \frac{\tau_F}{10} \quad \mu\,\text{watts/Å} \ldots. \qquad (3.1)$$

This represents the flux from one square degree of sky received by a six-inch diameter aperture at the bottom of the atmosphere.

In practice there was some noise on the calibration scan, so it was divided by the filter transmission, point by point, using a noiseless filter profile, and an average corrected count was found for the nominal 50-second integration time.

Putting in the instrumental parameters, experimental values, and those supplied by the NPL, for the narrow region around 5183.6 Å, the value of W arrived at was

$$W = 1.78\times10^{-9}\ \mu \text{ watts per Å per count.}$$

This may be conveniently expressed in photons. The energy E of a photon of frequency ν is given by

$$E = h\nu\text{, where } h\text{, Planck's constant, is approximately } 6.6261\times10^{-34}\, m^2 kg/\text{sec}.$$

From the wave equation,

$$c = \nu\lambda$$

where c is the speed of light, approximately 2.9979×10^{8} m/sec, and λ is in this case 5183.6Å, or 5183.6×10^{-10} m.

Thus the number of photons represented by W is

$$\frac{W}{E} = \frac{1.78\times10^{-9}\times10^{-6}\times5183.6\times10^{-10}}{6.6261\times10^{-34}\times2.9979\times10^{8}}$$

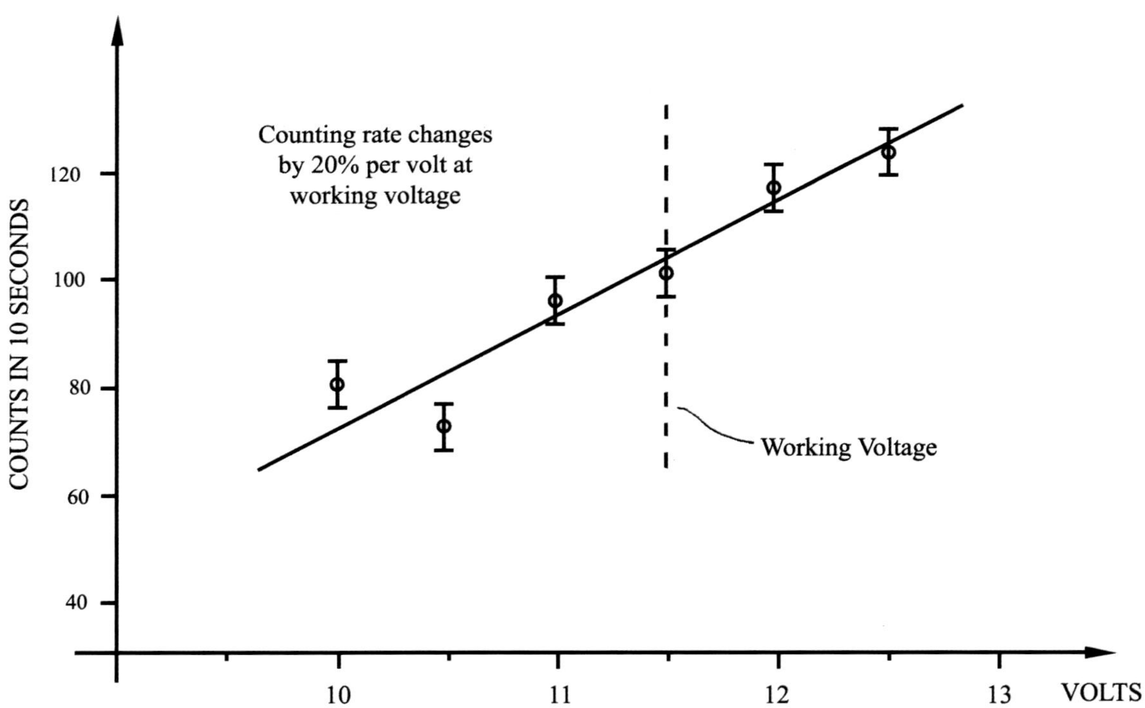

Figure 3.2 Experimental graph of counting rate against voltage across bulb for the intensity calibration set-up.

Hence 1 count represents 4640 photons/Å .... (3.2)

It is sometimes useful to know how many photons are arriving in total for a given count rate. The equivalent width of the F-P pass-band was typically 1.5 Å, so

1 count represents 6960 photons .... (3.3)

3.1.2 Accuracy of intensity calibration

Figure 3.2 (previous page) shows a graph of counting rates measured in the laboratory using the lamp at different voltages. An uncertainty in the AVO meter reading of 0.1 volts produces an uncertainty in counting rate of only 2 per cent. The National Physical Laboratory's calibration was to an accuracy of 2 per cent also, so the overall accuracy is limited only by photon shot-noise, small temperature fluctuations of the lamp, geometrical uncertainties and the estimation of the transmission of the coelostat. For sources outside the atmosphere, extinction losses also contribute to the uncertainty. The final calibration is believed to be accurate to a factor of 4.

3.2 Wavelength calibration

The aim of wavelength calibration is to assign a true wavelength to every programmed pressure point found by the pressure controller. Since in our case each programmed point corresponded to the positioning of a rotating arm illuminating a particular pea-lamp (see Hindle and Reay 1967), the programmed point scale is referred to by 'lamp number', L. The calibration involves (a) determination of the unshifted MgI 5183.62 Å position for each scan and (b) determination of the separation of this line and the Zinc 5181.99 Å line in terms of 'lamp number'. From this can be calculated the wavelength interval between each programmed point (in 'angstroms per lamp'). We will see that this quantity $d\lambda/dL$ may vary slightly with lamp number itself, because of the non-linearity of the pressure controller.

3.2.1 Determination of the datum line position

For the purpose of determining the 5183 Å position, celestial sources (planets, Moon) were considered unsuitable because of their lack of availability near the time of a scan, and the difficulty of evenly illuminating the entrance aperture from a small source on the sky. Instead a Magnesium lamp was used. Because of the slight drift of the etalon over the period of a scan, the lamp was scanned immediately before and after each scan, and the datum line position was determined as the mean value of the two measurements.

Some early calibrations were done using a stack of diffusers illuminated quite closely by the lamp. At the time, the arrangement was considered adequate, but soon it was suspected that the illumination of the diffuser was slightly greater at the centre than at the edges. To test if this was so, scans of this source were compared with scans of the same lamp using the more evenly illuminated diffuser described in Chapter 2. The test involved measurement of a differential of only about half a 'lamp' in a situation where the drift of the peak from scan to scan and even during one scan was of the same order. A 'dynamic' method was adopted, in which the drift and the differential could be easily identified, and measured separately. The technique was to align the etalon, and then make a number of scans alternately using the two configurations, without realignment. A typical result is shown (Figure 3.3),

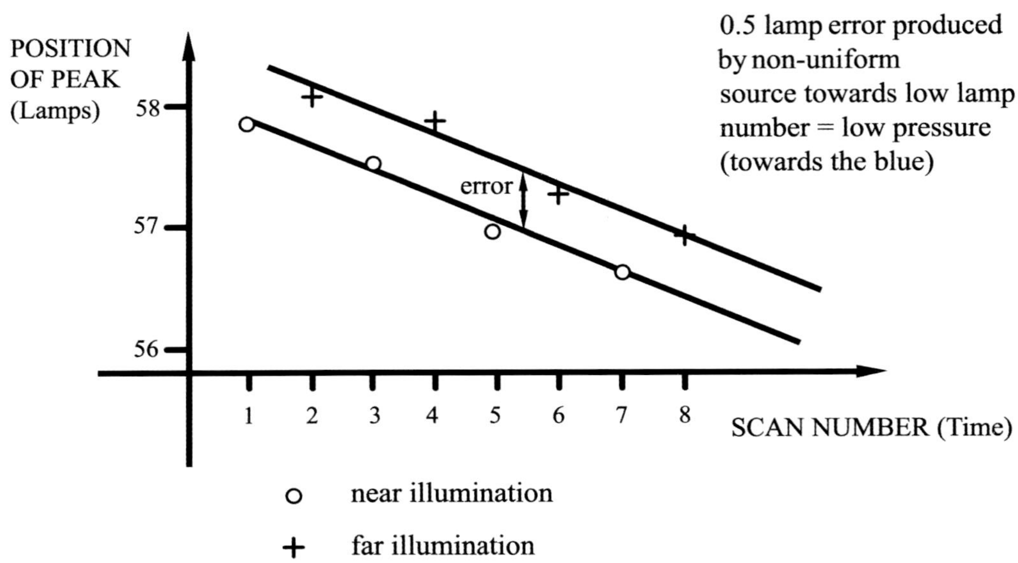

Figure 3.3 The drift method applied to determination of error due to non-uniform illumination of entrance aperture.

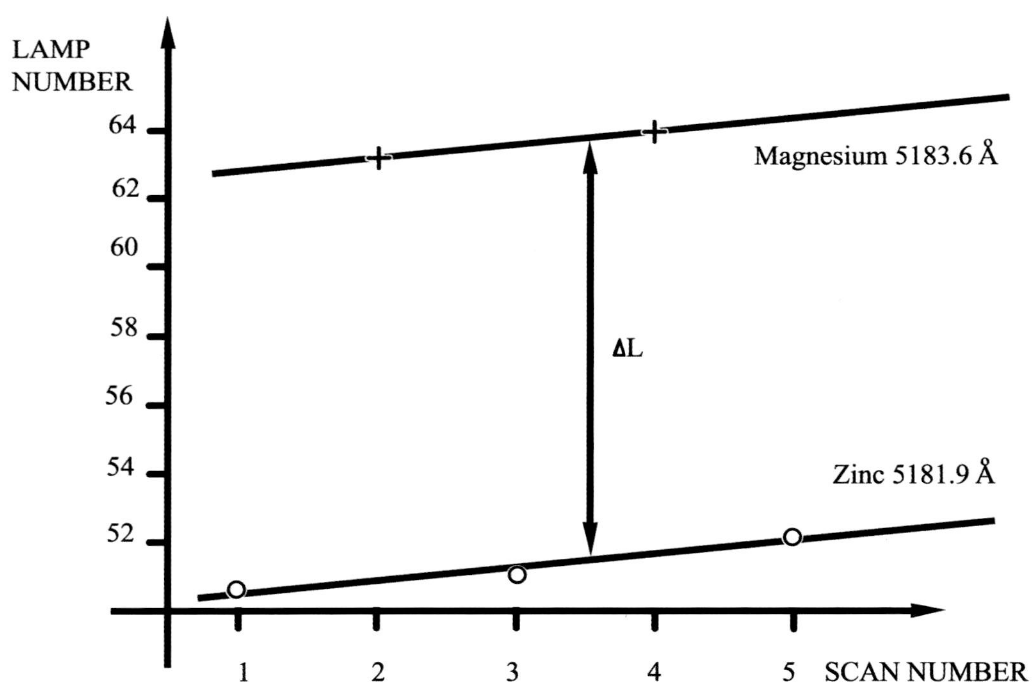

Figure 3.4 Determination of the separation of the two calibration emission lines using the drift method.

plotted as a function of time. Lines joining successive determinations of any one configuration clearly show the drift of the etalon. The separation of the two lines at any time gives an accurate measure of the error in the old configuration, which was found indeed to be 0.5 lamps, to the blue, as would be expected. This correction was applied to the early calibrations, and all subsequent calibrations were performed using the new uniform diffuser for illumination.

3.2.2 Establishment of wavelength scale

The separation in lamp number of the MgI and Zn lines ΔL was found by a similar method – i.e. scanning the two lamps alternately as long as the finesse stayed good; the positions of the peaks were plotted against time, and the separation at an instant measured from the graph (Figure 3.4).

In this case the two lines appear at different wavelength positions under the filter profile and it would seem at first sight that the filter should be divided out to give the true positions. However the shape of the maxima is entirely instrumental in origin, and arises from a line of virtually zero width, so the shape of the filter profile could have no effect on the position of the peak. Strictly speaking it is only correct to divide out the filter after deconvolving the instrumental function from the spectrum. In the case of a monochromatic source this reduces the spectrum to a line. However the sky spectra under study contain no important narrow features, so the filter division process is valid for these.

With all corrections made, the values of ΔL were plotted against the position in lamp number of the Zn line (Figure 3.5), which varied depending on how the etalon was aligned at the time.

A linear relationship between lamp number and pressure would give simply the straight line ΔL = constant, but experimentally this was not found to be the case, indicating a non-linearity in the pressure

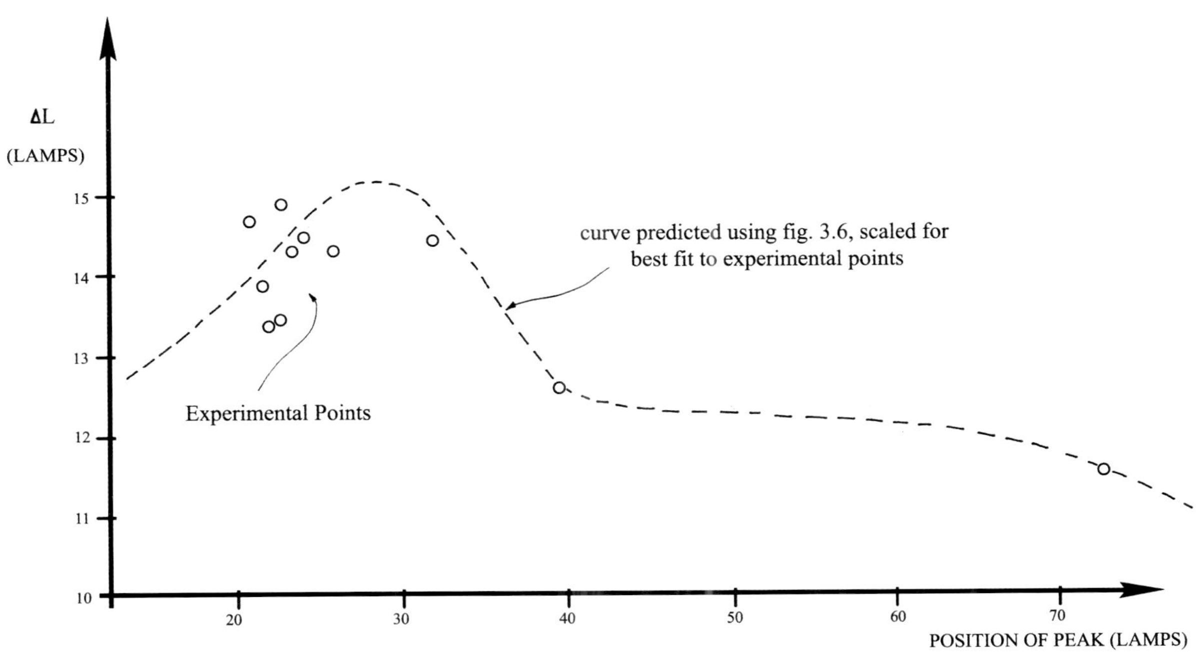

Figure 3.5 Experimental graph of separation of Zinc and Magnesium lamps, as position of Zn peak is varied.

controller. A check was made on readings of the Bourdon pressure gauge for each programmed point and a non-linearity was confirmed (Figure 3.6). For one section of its movement, the controller gave less pressure variation per lamp than expected – around $L_{Zn} = 20$.

The part least affected was around $L_{Zn} \approx 40.5$, so this was chosen for future sky scans, and is a convenient point to evaluate ΔL in terms of wavelength.

For $L_{Zn} \approx 40.5$, the measured value of ΔL was 12.9 lamps. Referring to the pressure/lamps graph (Figure 3.6), at $L = 40.5$, a ΔL of 12.9 corresponds to a pressure difference of 10.0 lbs per square inch. It is this pressure difference which is a constant for a given wavelength interval, so going back to Figure 3.6, we can predict what ΔL will be for any position of the Zn line. The curve so obtained is shown dotted in Figure 3.5 a good fit to the experimental data. Thus a value of $d\lambda/dL$ can be confidently adopted for each position on the scale of lamps. The values adopted were:

$$\text{For } L < 30 \qquad d\lambda/dL = .134 \text{ Å per lamp} \qquad \ldots\ldots\ldots \qquad (3.4)$$

$$\text{For } 30 \leq L \leq 41 \qquad d\lambda/dL = .102 \text{ Å per lamp} \qquad \ldots\ldots\ldots \qquad (3.5)$$

$$\text{For } 41 \leq L \leq 95 \qquad d\lambda/dL = .136 \text{ Å per lamp} \qquad \ldots\ldots\ldots \qquad (3.6)$$

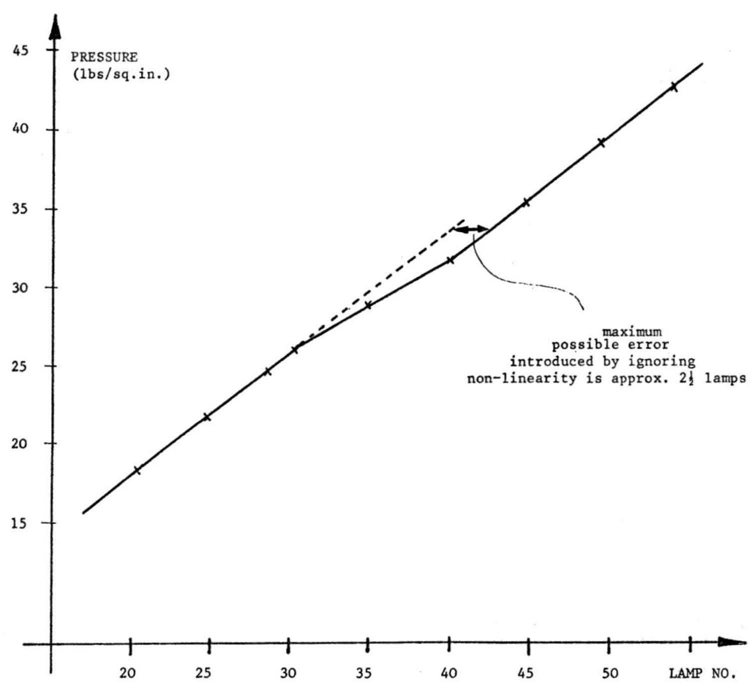

Figure 3.6 Experimental variation of pressure as measured by Bourdon gauge against pressure controller position in lamp number.

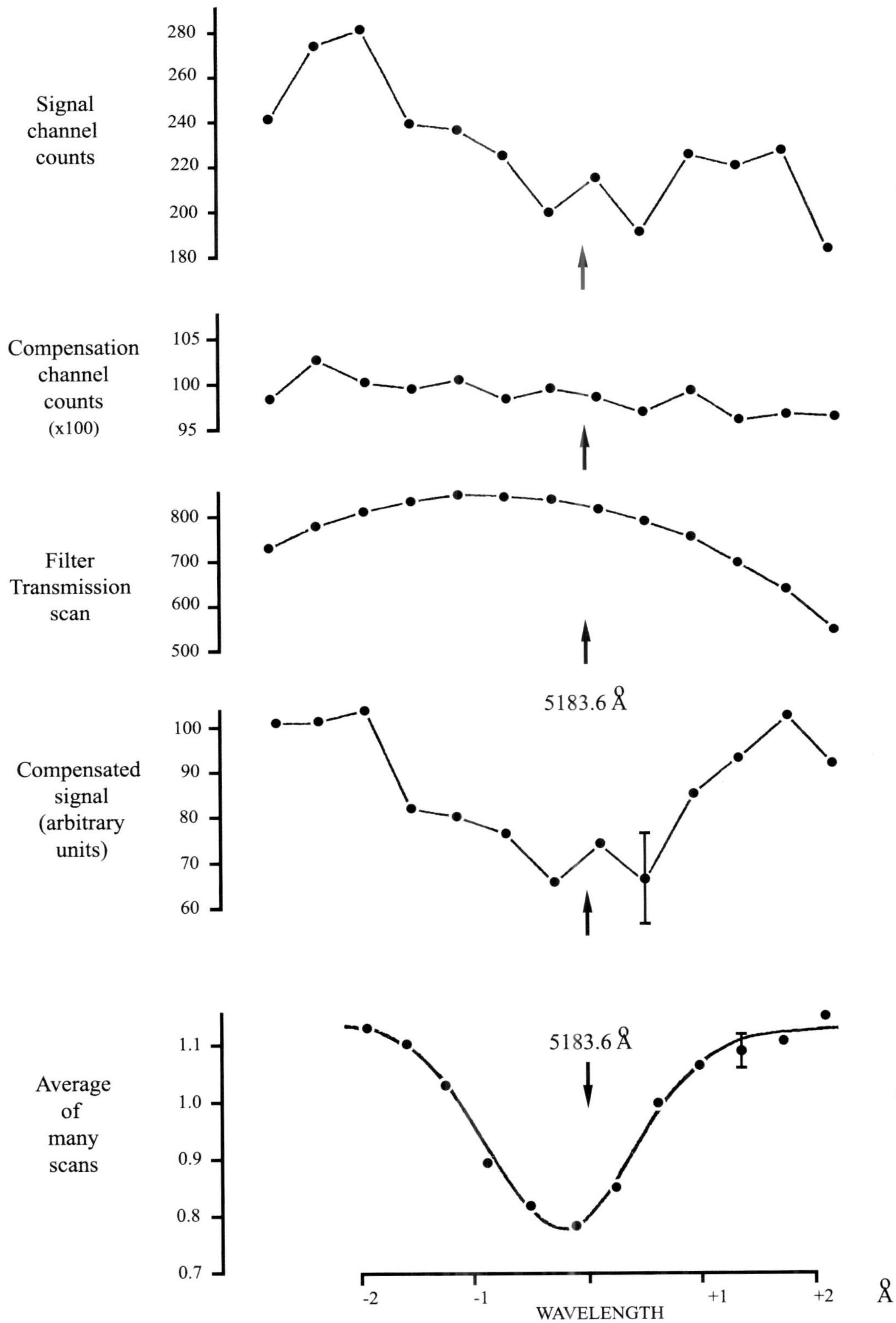

Figure 3.7 Illustrating the treatment of the data.

Hence a wavelength scale was established for all scans. The accuracy of the scale is the accuracy to which the dotted curve in Figure 3.5 can be fitted to the experimental points – i.e. about 0.6 lamps, corresponding to between 0.06 and 0.8 Å. The overall accuracy of the position of features in the final corrected spectra is dealt with in section 4.4.

3.3 Treatment of a single scan

A typical 'raw data' scan is shown in Figure 3.7a (previous page), which is plotted directly from the printout values of the signal channel. In the diagram the scale of lamp number has already been converted to a wavelength scale, as was done for each scan using the results of section 3.2. The arrow indicates the position of 5183.6 Å as determined from the two associated finesse scans of the Magnesium lamp, before and after.

Figure 3.7b shows the corresponding compensation channel record, and Figure 3.7c plots the filter shape, an average of scans of a tungsten source performed at the beginning and end of the night's observing. Starting with data (a), the dark count ($\approx$60) was subtracted from each point, and the resulting true signal count divided by (b) and by (c), giving the required compensation in overall light level. The result is shown in Figure 3.7d. Figure 3.7e shows a typical averaged spectrum, the result of combining the data from several scans of the same region of sky (discussed in Chapter 4).

The spectra were multiplied by a constant in each case, which was adapted to produce a 'normalisation' taking into account the various changes in 'comp' channel geometry and fore-optics during the experiments. This enabled a direct comparison to be made between the intensities of scans taken at different times, sometimes showing up unusual atmospheric conditions. The fall-off at the edges of the raw data scan is seen to have been entirely due to the interference filter, and the Poissonian error bars are consequently large for these 'edge' points.

The whole process of plotting the two filter scans for each night with their associated finesse scans, lining up the calibration positions and averaging the two scans, and then determining from this the correct filter transmission to be applied to each point in every scan, was laboriously done by hand for the October data, but for the April results the process was streamlined using the experience already gained of the instrument's behaviour, and a computer program written by T. R. Hicks was used for the reduction process.

Each scan was checked against comments written at the time of the scan for any anomalies which could be assigned to local interference, or obvious failure of the counting system. In spite of all precautions (Chapter 2), occasionally spikes in the electrical supply lines, or aerial pickup from the nearby radio transmitter, caused the counters to spontaneously reset. The scans were scrutinised for such effects, and in the event of a count being lost, the chart recorder record was used to estimate a true count for the single point affected. Only a small degradation of signal accuracy was produced by this process. The scans were inspected for any obvious asymmetry in continuum levels. Where this could be traced to a probable misplacement of the filter in dividing out, a better estimate was made, and the scan re-computed. Sometimes it was concluded that asymmetries were due to other causes (pre-saturation of the signal or comp channel during a calibration scan, contamination by clouds, stars, Milky Way or twilight, abnormal finesse drifts or degenerations). Separate corrections for these were made where possible.

3.4 Daylight correction

In the case of scans recorded near dawn or dusk, a new method was adopted to compensate for daylight contamination. In morning observations, the onset of twilight (dawn) within 10 degrees of the horizon, where much of the work at low elongations was done, did not always coincide with the beginning of astronomical twilight (calculated as the time when a seventh magnitude star should be just visible in the Zenith); it varied as much as half an hour depending on the local atmospheric conditions towards the East. Similar comments apply to the end of twilight at the beginning of the night. A scan recorded at dawn is shown in Figure 3.8. Part (a) shows the signal counts recorded, and in part (b), the compensation channel record, the beginning of daylight contamination, as opposed to the slight slope cause by the rising Zodiacal Light cone, is clearly visible, and it is reasonable to suppose that whole of the shaded portion is daylight. Scans of sunlight were obtained during the day, and so this spectrum is well-known and virtually noiseless, owing to the high daytime fluxes, making it possible to reliably subtract it from the raw spectrum, to give a much improved final compensated scan.

If, at any spectral point, the fraction of daylight present is F(see Figure 3.8, overleaf), the appropriate correction to be subtracted to eliminate daylight is $F \times D(\lambda) \times C$, where $D(\lambda)$ is the daylight spectrum scaled to unity continuum, and C is the assumed true continuum level in the spectrum under study. The function $F \times D(\lambda) \times C$ is shown in Figure 3.8c and the corrected spectrum in Figure 3.8d. Apart from restoring the balance of the continuum levels, the correction 'fills in' a broad area near zero shift position, since a depression in continuum is being 'subtracted' here. The small magnitude of the corrections in the important central region indicates that the method could be used to good effect even further into twilight, and at even smaller elongations. A small amount of extra noise is introduced owing to the fact that irregularities in the compensation channel may be due to variations either in ZL flux, or the daylight flux, or both. Thus the error bars towards the high wavelength side of Figure 3.8d are larger than normal.

At the stage when each individual scan had been fully reduced, a preliminary examination of them immediately revealed the existence of what appeared to be small *emission* component at the MgI wavelength, variable in intensity.

This emission is treated here first, as part of the reduction section, since an understanding of its relevance is necessary before proceeding to the determination of radial velocities in the dust cloud; the rest of this chapter is devoted to the MgI 'emission core'.

3.5 MgI emission

3.5.1 Discovery

In describing the central feature, the word 'core' is used because, whereas the main absorption line is roughly the width of the solar Fraunhofer line, slightly widened by the instrumental profile - about

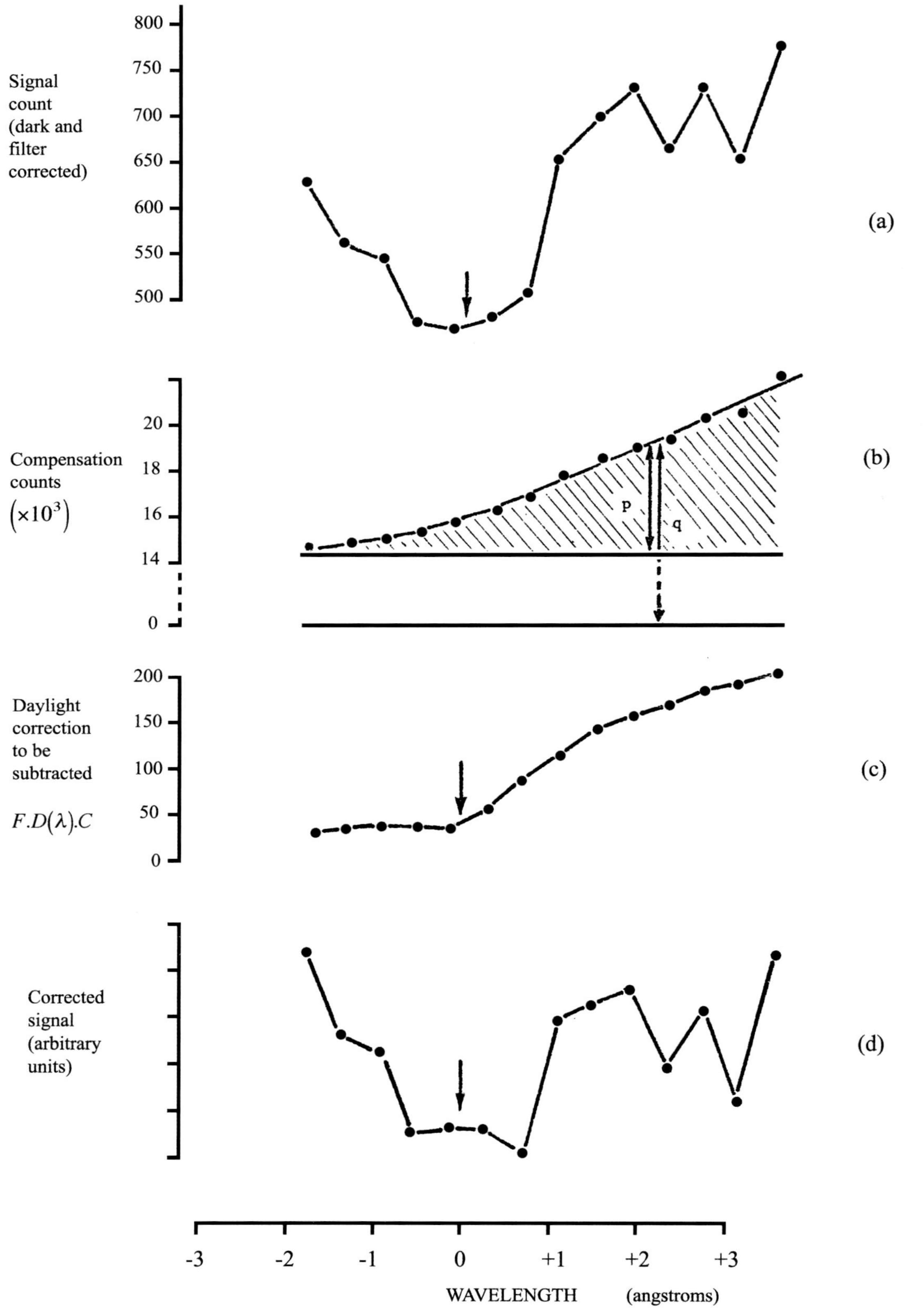

Figure 3.8 Correction for daylight contamination around dawn.
The fraction of daylight present at any point in the spectrum is p/q in (b).

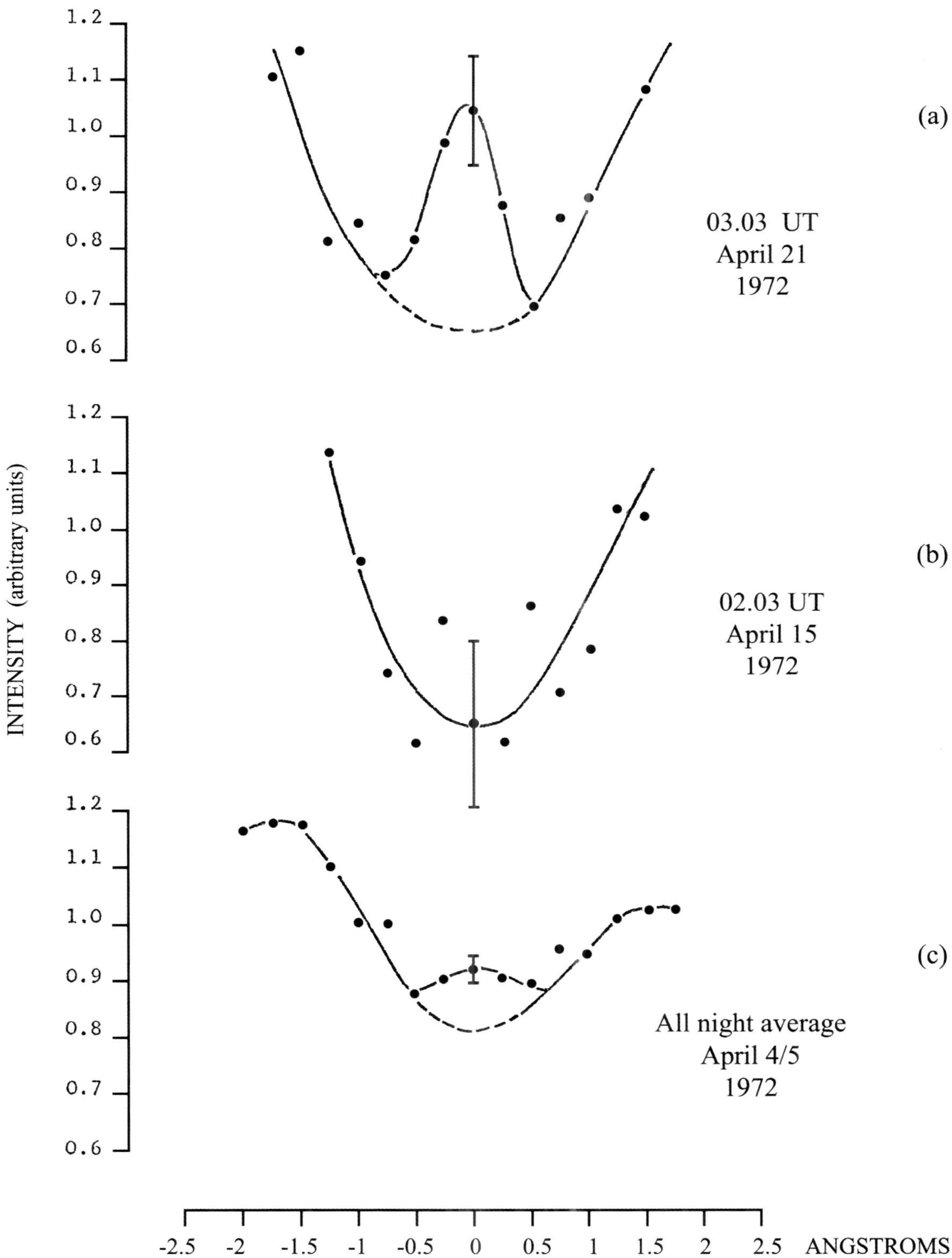

Figure 3.9 Evidence for emission in the data.

3 Å in all – the emissions generally have widths of the order of the instrumental resolved width itself – about 1.5 Å. They appear as a narrow peak inside a trough. An example is shown in Figure 3.9a (previous page). Figure 3.9b shows a scan in which there is apparently no emission.

In general, such individual scans were not of good enough quality to make measurements from, but for some of the data procured in adjacent parts of the sky on the same night it was possible to construct average scans which yielded good quality data (Figure 3.9c), and in some of these the core appeared to be real. The method of averaging was applied in a different way for investigating the absorption feature itself, and is described in full later (section 3.6).

It was at first thought that the core could be the result of seeing two absorption lines at a narrow separation, giving a double feature. This can easily be shown *not* to be a possible explanation, however. Starting with the solar Fraunhofer line and working from models of co-and contra-rotating dust in solar orbits, it is possible to produce a double feature with approximately the required spacing, but the central core turns out to be too far from the zero-shift position to fit our data, which always show the core at zero position. Furthermore, when the theoretical spectrum is convolved with an instrumental function one and a half angstroms wide, as exhibited by our instrument, the double feature completely disappears, replaced by a single absorption feature with a slightly greater width. It is therefore with some confidence that we believe the core to be due to emission in the Earth's atmosphere.

3.5.2 Wavelength of the MgI emission

In the individual scans the wavelength of the emission peak could be estimated to better than 0.2 Å from those scans which clearly showed a core. By combining the measurements from all such scans the emission wavelength was determined to be 5183.6 Å $\pm$ 0.1 Å. Reference to the MIT wavelength tables shows that lines of Cobalt, Europium, Zirconium and Magnesium occur very near to 5183.6 Å. We identify the emission as the 5183.6 Å line resulting from the $3p^3P_2 - 4s^3S$ transition of MgI.

3.5.3 Intensity of the emission

On each scan an estimate of the height of the emission peak was made by interpolating the probable uncontaminated absorption feature, and measuring the central discrepancy, in normalised scale units (see section 3.3). To convert this measure into intensity it is first necessary to 'un-correct' for the compensation channel division (an average value for the scan was used for this), and for the normalising constant, to get back to the original counts, corrected only for filter transmission.

To obtain the absolute value of the number of photons arriving at the bottom of the atmosphere, the intensity calibration (equations 3.2 and 3.3) was applied, and the results subsequently converted to Rayleighs. As an example, in the October results, the overall average of the average nightly counts recorded in the sample period of 48 seconds estimated to be due to emission was 17 counts.

Now since one count represents 6960 photons (equation 3.3), the number of photons received in sample period was 118,320. The area of the six-inch entrance aperture is

$$\frac{\pi}{4}(6\times2.54)^2 \text{ sq cm} = 175 \text{ sq cm}$$

Therefore number of photons per square centimetre in 1 second is approximately

$$\frac{118320}{175\times48} \approx 14$$

The instrument accepts a solid angle of 0.00056 steradians. Thus the number of Rayleighs is

$$\frac{14.08}{5.6\times10^{-4}}\times\frac{4\pi}{10^6} \approx 0.32 \text{ Rayleighs.}$$

0.32 Rayleighs is thus a typical MgI emission rate as observed. Measurements in this wavelength region give extinctions of typically 0.15 magnitudes in the zenith for the Izaña site.

Figure 3.10 (overleaf) shows the average counting rates for each night for the periods of the observations.

3.5.4 Variation of daily average emission

The well-established enhancement of CaII emission at times of high meteor activity (Vallance and Jones 1956) suggested to us that there might be a correlation between MgI emission and meteor showers. The Orionid shower has a maximum activity around October 20th and the Lyrid shower about April 21st. The September and October 1971 data show little significant variation in emission from night to night, having a mean level of 0.31 Rayleighs. The April 1972 data show a generally lower emission rate of about 0.1 Rayleighs, with some interesting structure beginning on the night of April 14/15, coinciding with the onset of the Lyrid meteor shower.

Two scans recorded near morning twilight on April 15th showed an unusually strong emission of about 0.7 Rayleighs. This declined over the next few nights until another enhancement occurred on April 20/21, corresponding to the time of maximum activity of the Lyrid shower. Unfortunately our data do not extend beyond April 21st. It is worth noting that the 'enhancements' in late April 1972 merely bring the emission rate up to the September-October 1971 level. In this respect it is not clear whether the emission really underwent an enhancement in late April or is merely in a depressed state in early April! We believe the enhancement to be real, but the difference in ambient emission rates September-October 1971 and April 1972 may be an artefact of the calibration technique, and modifications to the instrument's fore-optics between observation periods, though every effort was made to correct for these factors.

Enhancement of the emission at the time of the Lyrid shower suggests that at least part of the atmospheric magnesium is of meteoric origin. No such enhancement occurred during the October 1971 Orionid shower, however.

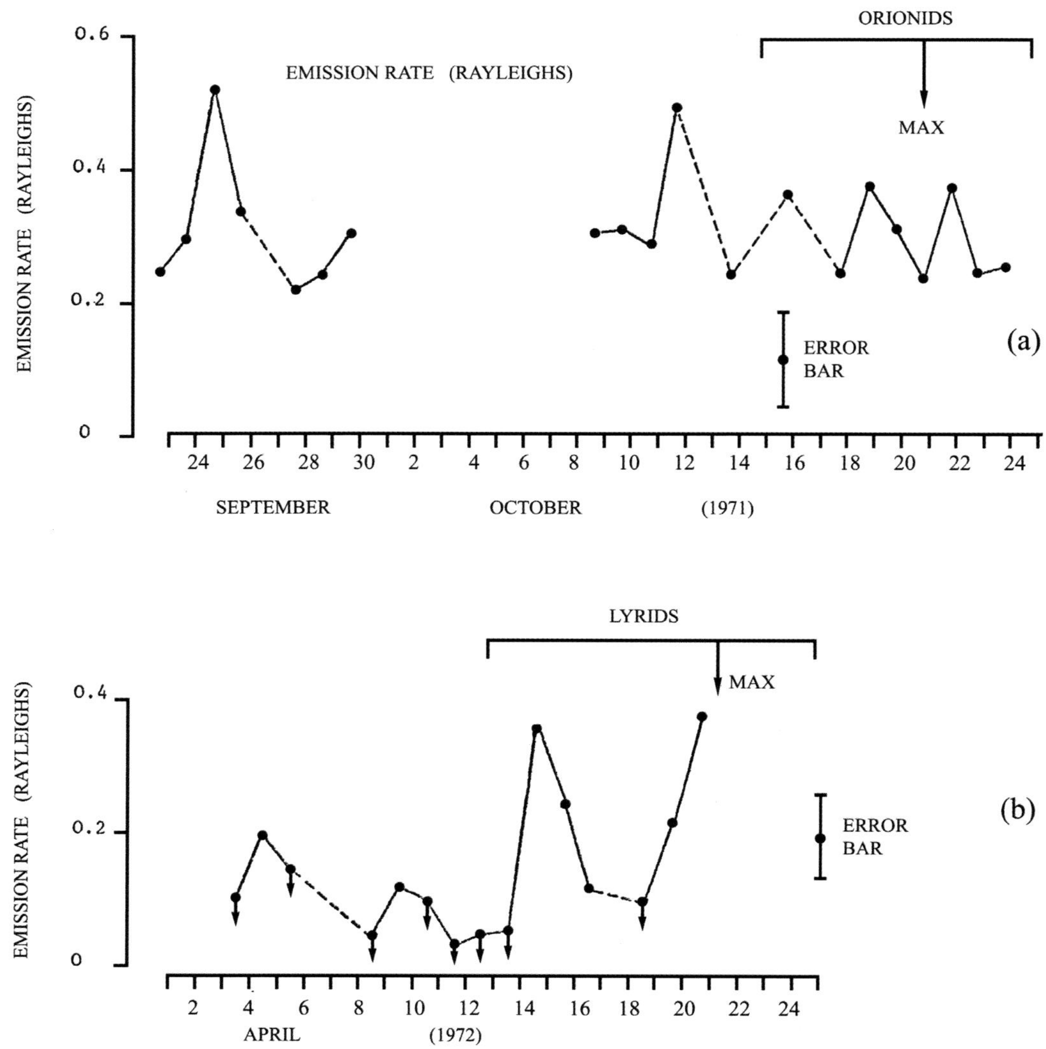

Figure 3.10 Mean night-time emission as a function of time. The vertical arrows indicate upper limits, and the gap in data corresponds to the period of Full Moon.

3.5.5 Variation of emission throughout the night

Depending on which mechanism produces the Magnesium emission, different variations of intensity throughout a night could result, so an attempt was made to determine from the data an hourly estimate of intensity averaged over a number of nights. The process was attempted only for the September-October period, since a greater number of scans were affected.

The result is shown in Figure 3.11. The error bar has been calculated from the probable errors in the individual estimates assuming a normal distribution of these about a mean value. Each point represents the average of about 20 measurements from individual scans. It is seen that there is a significant increase in intensity near twilight in morning and evening. There is present in the observations, however, a strong observational selection effect. Most of the scans recorded near sunset or sunrise were at low altitude, and airglow phenomena are known to increase in brightness near the horizon (the Van Rijn effect) so

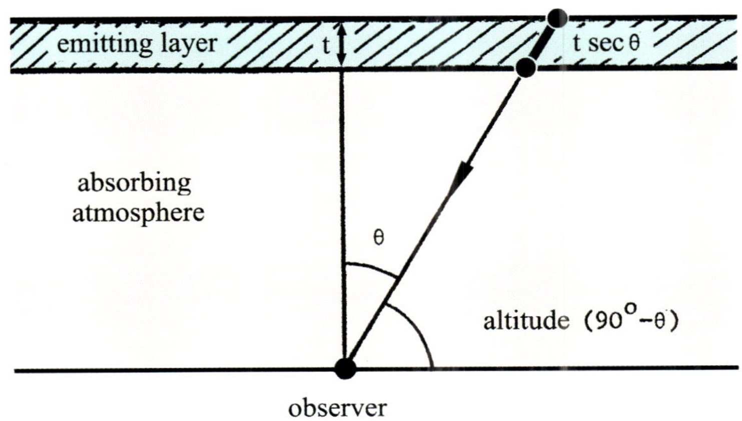

Figure 3.11 Hourly averages of MgI emission intensity throughout the night.

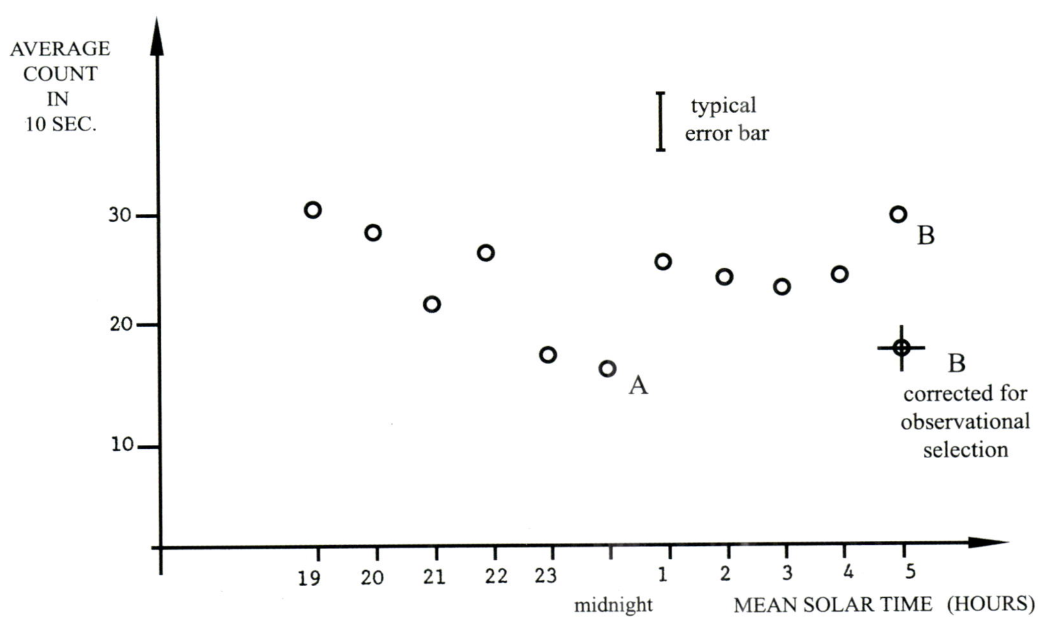

Figure 3.12 Simple model of an airglow layer.

these scans might be expected to show a greater intensity of emission than those around midnight. A simple model of an airglow effect is shown in Figure 3.12. If the emitting layer is optically thin, the total intensity of emission recorded on the ground is proportional to the line-of-sight length of path in the emitting layer, and thus …

$$\text{Emission} \propto \sec\theta \text{, where } \theta \text{ is the zenith distance.}$$

The length of absorbing path is also roughly proportional to $\sec\theta$, but more precisely is proportional to $f(\theta)$, a function defined using the air mass figures tabulated in Allen's Astrophysical Quantities (paragraph 54). The intensity transmitted is thus

$A\sec\theta \times e^{-kf}$ where k is a constant for a given air type, and A a constant governed by the physical mechanism of the emission, and determinable by experiment. If kf is small ...

$$\text{intensity} = A\sec\theta(1 - kf + \frac{kf^2}{2})$$

The approximation (kf small) is good only as long as the minimum transmission is about 80%, which is just realised at the lowest elongations under discussion. Using the tabulated values, the relative strengths of emission at two different altitudes can be found.

To predict what the effect of observational selection might have been, the scans providing the information for points A and B in Figure 3.11 were individually checked for zenith distance. For each scan the relative strengths from a typical airglow distribution were calculated as above, and the average values were adjusted accordingly. The result was that, to compensate for the observational selection effect due to zenith distance, the value at point B needed to be multiplied by 1.44. As can be seen, this brings point B to roughly the same level as point A, and suggests that the true variation of intensity throughout a night is small.

3.5.6 Final comments on MgI emission

The fact that the MgI emission does not disappear late in the night when the 90-100 km region of the Earth's atmosphere, where most other metallic ions are observed (Narcisi 1968; Narcisi and Bailey 1965) is completely in shadow, makes it unlikely that any fluorescence mechanism is a source, and indeed it is hard to see how fluorescence could excite the triplet $3p^3P_2 - 4s^3S$ transition. Radiative recombination of Mg+ with an electron, the mechanism suggested by Anderson and Bath (1971) governing the daytime balance between Mg and Mg+, would seem to be the most likely cause of the emission. It might be expected that other Mg lines, such as the other lines of the triplet, 5172.68 Å, and 5167.33 Å, would be present also. The interecombination line at 4571.1 Å connecting the 3p to the 1s state should also be present. I know of no observational searches for these lines, though more recent studies seem to show that neutral Mg emission is highly variable over a day, with a maximum at dusk (Gardner *et al* 1995).

The discovery of this MgI emission was the subject of the first Hicks, May, and Reay paper, published in Nature in 1972, reproduced in Appendix 5.

Attention was now turned back to the absorption line, at 5183.62 Å, for the purposes of determining Doppler shifts as originally planned, and for this purpose the small emission core, once subtracted, was ignored.

3.6 Averaging

The process of using the observed absorption line spectra to discriminate between theoretical dust cloud models is complex. Even if we confine our attentions to a ZL produced exclusively by a circumsolar cloud of material, with no other components, there are still many variables. The spectral profile of the absorption line may be expected to depend upon elongation from the Sun, and possibly the position of the Sun on the celestial sphere (reflecting the Earth's position with respect to the Zodiacal Cloud). There may also be short, medium and long period variations due to solar activity, or due to changes in the density and composition of the cloud with time, perhaps brought about by the visits of comets to the inner Solar System. Because of the way in which the data was collected, it was considered practical to approach the analysis purely as a function of elongation, but to keep the two sets of observations (from 1971 and 1972) separate, in order to highlight any differences between the two periods.

The ecliptic plane was divided into 10-degree intervals (or 5 degrees for low elongations), and an *average* spectrum was computed for each group of scans in the interval. The method was to set up a number of wavelength 'channels' of width one 'lamp', and for each channel, add up the relevant number of counts recorded for each scan, finally dividing the total by the number of scans. Care was taken to avoid introducing artificial noise by averaging over a different number of scans for different points. Where no value existed in a given channel, an interpolated value was used. Because of the finite width of a channel, this method produced a final spectrum which has been convolved with a 'box-car' function, one lamp in width. However the effect of this is quite negligible compared with that of the instrumental function width.

A typical averaged scan, in this case an average of 8 individual scans, is shown in Figure 3.7e. Most of these derived spectra exhibit what we might call a 'well-behaved' absorption feature. All averages were done on the central four-and-a-half angstroms of the scans only. Attention was then turned towards fitting curves to the experimental points, in order to extract parameters which can be compared with theory. I attempted first to fit simple polynomial shapes to the points.

3.7 Polynomial fits

3.7.1 General

In 1973, the sole computing power available at Imperial College for processing data was contained in a single 'main frame' Atlas computer, housed in the Mechanical Engineering building, and taking up a whole floor of space. Input was by means of punched cards or 5-hole paper tape, and due to the demand, and the slow processing speed, it was normally possible to run a programme just once or twice a day. It was thus a very time-consuming business to 'debug' a new program, and much patience was needed. It was possible to output on a printer, but the choice could be made to also output on card or paper tape media, and, in this form, the result could be used directly as input to a subsequent program. This was the approach I used, producing decks of cards of the averaged scans and fitted curves for input later to a graph-plotting programme. Beginning with no assumptions about the shape of the line, I computed a best-fit polynomial curve for each averaged scan. The computer program used, LSQFIT, was an adaptation of a CERNUP library program which fits, by the method of orthogonal polynomials,

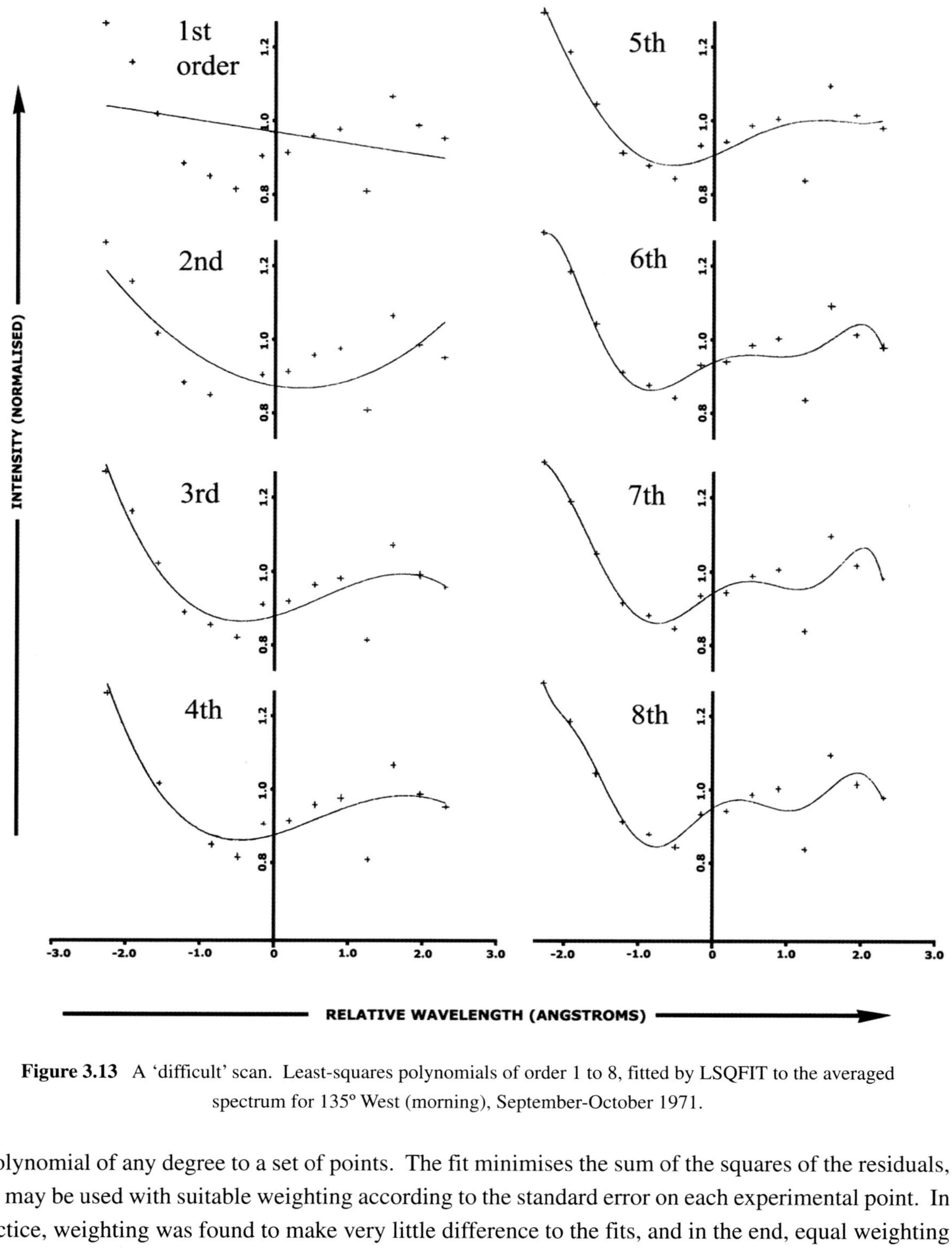

Figure 3.13 A 'difficult' scan. Least-squares polynomials of order 1 to 8, fitted by LSQFIT to the averaged spectrum for 135° West (morning), September-October 1971.

a polynomial of any degree to a set of points. The fit minimises the sum of the squares of the residuals, and may be used with suitable weighting according to the standard error on each experimental point. In practice, weighting was found to make very little difference to the fits, and in the end, equal weighting was used in all cases.

Depending on input format, the program printed out various information on the fits. One of the most useful is the *"F-Ratio"*, which gives a measure of how good a fit a polynomial of a given order is, relative to the that of a polynomial of the next order up. As would be expected from the shape of the curves, in most cases a fourth order polynomial gave a good fit (three possible turning points). Figure 3.13 shows a scan for which it was particularly hard to find simple fit; it provides a good illustration of the way the shape of the curve can vary with successive orders of the polynomial.

Figure 3.13, as others in this report, is from a Calcomp plot. Program LSQFIT was equipped with a card output giving the 'x' and 'y' values of the experimental points, and coefficients of the fitted polynomial. A program POLYPT was written to accept these as input, and plot the experimental points on the correct axes, and to trace out the fitted curve, calculated at one-twentieth angstrom intervals from the coefficients. (Printouts of the programs used are shown in the Appendices at the end of this thesis). These curves, fitted to the data, give an immediate eyeball estimate of the shift exhibited by the absorption line profile, as compared with the profile in direct sunlight. But, to remove human error from these estimates, methods were devised to extract an *average shift* parameter in each case. In this context, the word 'average' means a best estimate of the shift of the line profile. From Zodiacal Cloud models, as we shall see, a corresponding 'average shift' can be computed, the average of the shifts of the various 'cells' of dust particles along the line of sight integral for a particular elongation, for comparison with this *observational* 'average shift' (see Figure 1.3). I used two separate methods for extracting a best estimate of the shift from the fitted polynomial curves:

1) A determination of the minimum point of the curves (program MINIPT).
2) A determination of the wavelength half-way between the wavelengths of the two inflexion points (program FLEXPT).

It would also be possible to make an estimate of the shift by finding an ordinate which bisected the area contained between the spectrum cure and an estimated continuum level.

3.7.2 Computation of the minimum point (Program MINIPT)

Consider a fourth order polynomial. If the curve is represented by

$$I(\lambda) = \phi_0 + \phi_1\lambda + \phi_2\lambda^2 + \phi_3\lambda^3 + \phi_4\lambda^4$$

where λ is reckoned from a zero position at 5183.6 Å, and the ϕ's are the coefficients determined by LSQFIT, then

$$\frac{\partial I}{\partial \lambda} = \phi_1 + 2\phi_2\lambda + 3\phi_3\lambda^2 + 4\phi_4\lambda^3$$

which must be zero at the minimum point. Since all the $I(\lambda)$ curves have the general shape of an absorption line slightly shifted from zero position, we can assume that the turning point we want is the one closest to $\lambda = 0$. So any value close to zero can be used as a first approximation in an iterative process (Newton's method).

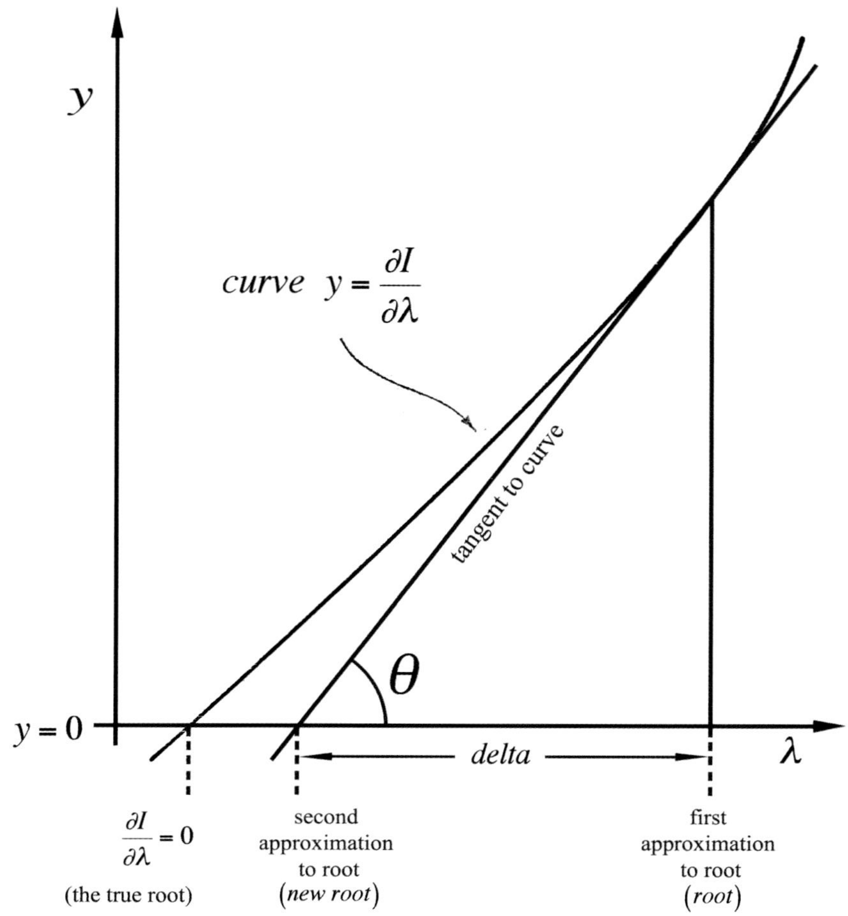

Figure 3.14 Illustrating Newton's method of successive approximation to solve the equation $\partial I/\partial\lambda = 0$, giving the minimum point of $I(\lambda)$.

Referring to Figure 3.14, given any first approximation *'root'* to the true root, we can find a better approximation by finding where the tangent to the curve at this point crosses the zero line.

$$\tan\theta = \frac{\partial^2 I}{\partial\lambda^2} = \frac{\partial I/\partial\lambda}{delta},$$

delta being the correction to be subtracted from *root* to give the value of '*new root*'.

Therefore

$$new\ root = root - \frac{\partial I/\partial\lambda}{\partial^2 I/\partial\lambda^2}$$

and the whole process is repeated, using *'new root'* as the new starting point, and so on, iteratively. We can stop the iterative process when $delta \leq 0.001\,\text{Å}$.

Now, differentiating again,

$$\frac{\partial^2 I}{\partial \lambda^2} = 2\phi_2 + 6\phi_3\lambda + 12\phi_4\lambda^2$$

So, using these values, the minimum point can be found for each curve, the distance of these from the zero position giving the shift directly. See Appendix 3 for the program MINIPT used to implement this operation. The figure below is included mainly as a reminder to myself!

3.7.3 Computation of the shift by the 'inflexion points' method (Program FLEXPT)

Again the analytical form of the 4th degree polynomial gives a simple expression for the shift measured by this method.

We require for an inflexion point

$$\frac{\partial^2 I}{\partial \lambda^2} = 0$$

i.e. $$2\phi_2 + 6\phi_3\lambda + 12\phi_4\lambda^2 = 0$$

This is a quadratic equation possessing roots λ_1 and λ_2, and the midpoint of the line joining the inflexion points has a λ value of half the sum of the roots.

i.e. $$\text{Shift} = \frac{\lambda_1 + \lambda_2}{2} = \frac{-(6\phi_3)}{2(12\phi_4)}$$

$$= \frac{\phi_3}{4\phi_4}.$$

Programs MINIPT and POLYPT were designed to read directly the output cards of LSQFIT, and the values of wavelength shift obtained were plotted against elongation from the Sun. The shifts computed from the minima in the 4th order polynomial fits are shown in Figure 3.15 (overleaf), separately for the two observing periods.

Since 4th order and 6th order fits were available for all averaged scans, the minimum point method was also applied to the 6th order fits, with less success because of the greater risk of ambiguity in the turning points.

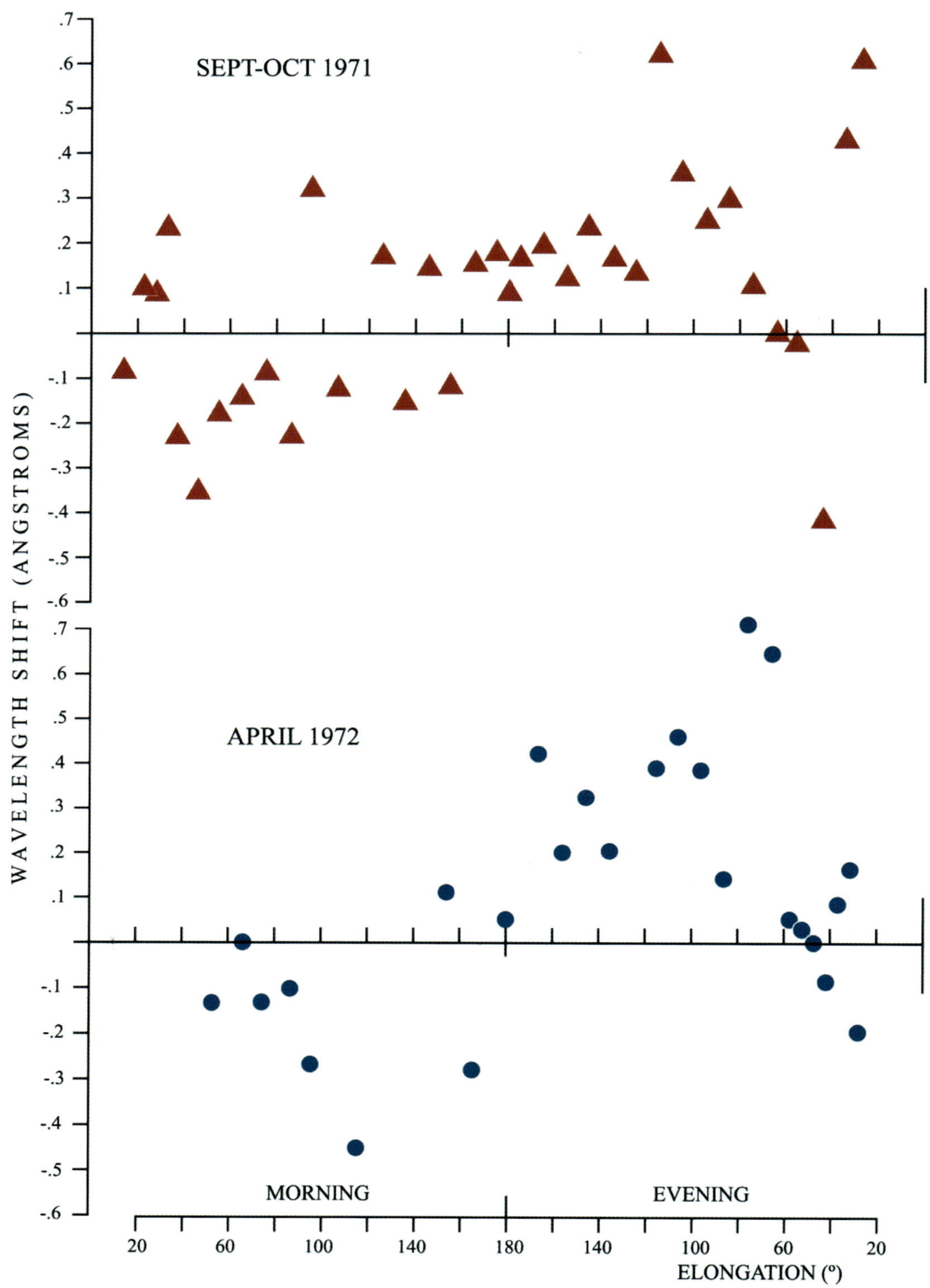

Figure 3.15 Wavelength shift estimates from polynomial fits. Plots of line shift vs. elongation using the minimum points of the 4th degree polynomials fitted to the averaged data as a measure of the shifted position. This method is considered unsatisfactory (see text), but it is pleasing to note that the general disposition of the points is quite similar to that found by the Gaussian fits method (see Figure 4.5).

3.7.4 Assessment of these methods

A necessary constraint on any method used to estimate the wavelength of the absorption feature is that it use the information from ALL the points in the optimum way, bearing in mind their various probable errors. It is also essential that the measure be identifiable with a physical quantity calculable from models of the dust cloud. In practice this quantity is the *'mean shift'*

$$\langle \Delta\lambda \rangle = \frac{\int \Delta\lambda . dB}{\int dB} \qquad \ldots\ldots\ldots \quad (3.7)$$

where $\Delta\lambda$ is the shift due to an element of brightness dB, and the integral is along a line of sight (Bandermann and Wolstencroft 1969; Reay 1969). If the original solar spectrum is symmetrical (a good approximation at 1.5 Å resolution) it is easy to show that $\Delta\lambda$ is also in this case identical to the shift of the bisector of the inflexion points, and to the minimum point.

Noise in a practical spectrum reduces the reliability of all three methods so far mentioned. The minimum point method depends critically on the accuracy of a few central points, so just a small emission feature, as was sometimes present, can be enough to upset the measure. The inflexion method is sensitive to the outer, more noisy, parts of the spectrum. The method of areas would rely on a good estimate of the continuum level, which again depends strongly on the outer points, which are noisy. A small error in the placing of the filter can produce an apparent asymmetry in the continuum levels to each side of the absorption feature. The results of the methods in practice bear out these comments. The scatter on them all is large, obscuring even the general trends visible to the eye in a casual inspection of the scans, and some results are obviously misleading if reference is made to the curves. For the final analysis, all of these methods were therefore rejected, except for the special cases (see section 4.1) where Gaussian fits could not be effected.

3.8 Gaussian curve fitting (Program GAUSSN)

3.8.1 Computation

The method finally adopted was to fit a Gaussian curve

$$I(\lambda) = I_1 - I_2 \exp\left[-k(\lambda - \lambda_m)^2\right]$$

where I_1 is identifiable with the continuum level, I_2 is a measure of the line depth, k is related to the width of the line, and λ_m is the wavelength of the central minimum point of the line. These four parameters correspond to the variables V_1, V_2, V_3, and V_4 in program GAUSSN, and in Hicks, May and Reay (1974). Once the Gaussian curve is fitted to the observations, λ_m is the most useful, since the difference between this and the unshifted wavelength, 5183.6 Angstroms, will give us the effective mean wavelength shift $\langle \Delta\lambda \rangle$ due to the Doppler effect.

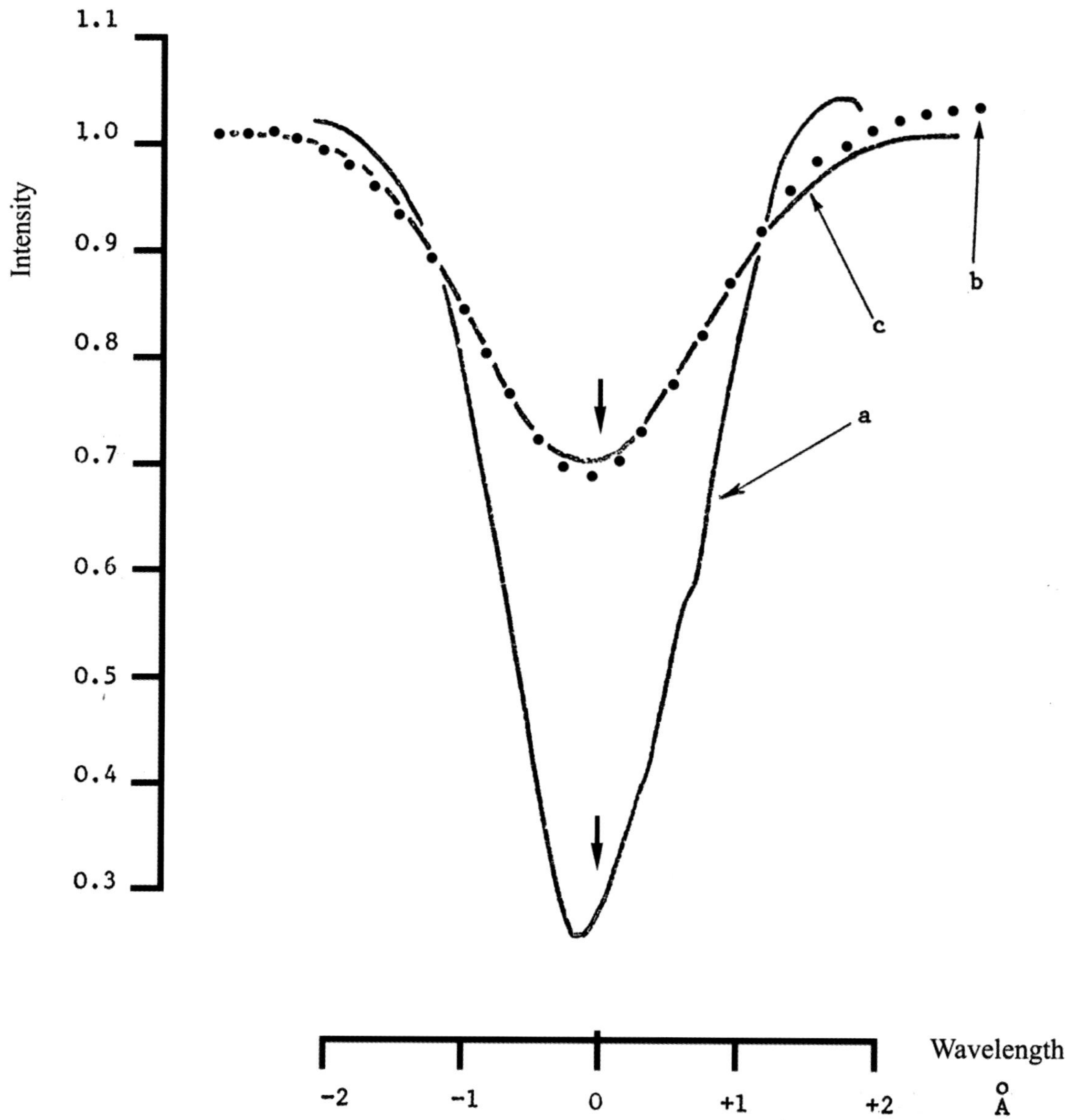

Figure 3.16 Comparison of a Gaussian curve with theoretical profiles. (a) A 30° elongation profile derived from a prograde dust cloud model with circular orbits. The dust is assumed to fall in number density away from the Sun inversely with radius of orbit. (b) This profile convolved with the instrumental profile of our Fabry-Perot. (c) A comparison Gaussian curve

An adaptation of a general least-squares fit subroutine was used to make program GAUSSN compute the fits. The program used an iterative method in the four dimensions, V_1, V_2, V_3, V_4, and worked from a first approximation, put in by inspection. It accepted directly the same card input format as LSQFIT. The output cards produced by GAUSSN were used by program GAUSPT to plot the points, and the fitted Gaussian curves.

3.8.2 Experimental case for the use of Gaussian fits

This method of Gaussian fitting is justified experimentally in that the fits obtained are not highly sensitive to the presence of a small emission core, nor to edge effects; also in most cases the essential residuals are small. Every experimental point contributes to the fit, and the resulting estimation of wavelength shift.

Further experimental endorsement of the method comes from the fact that the day scans performed with this instrument exhibit a shape which fits very well to a Gaussian curve.

3.8.3 Theoretical justification for the use of Gaussian fits

In order to simulate from theory the spectrum of the Zodiacal Light, I have used spectra computed from dust cloud models. The aim here is to show that by mathematically simulating the modifications made to a spectrum by the interferometer and subsequent reduction methods, a curve is arrived at which is very similar in form to a Gaussian. Curves derived from a variety of dust cloud models are discussed in detail in Chapter 4, but the method used in all cases is the line-of-sight integration through the dust cloud, described earlier and illustrated in Figure 1.3, in which it is assumed that there is a particle number density n with the dependence

$$n(a,r) = ka^{-p}r^{-2}$$

where a is the particle radius, r is the particle's distance from the Sun, and k is a constant.

For this test, the elongation was chosen to be 30 degrees and the dust was assumed to be moving in circular orbits in a prograde sense. A value of $p = 1$ was used.

The program CONVL (Appendix 4) read in the spectrum at one twenty-fifth angstrom intervals, interpolated to one fiftieth angstrom intervals, and assumed a continuum outside the absorption feature. It then read in the filter shape (approximated by a much wider Gaussian function) into an array, and shifted the array so that the peak was exactly at 5183.6 Å (equivalent to tilting the filter to 'tune' it in the experiment). The input spectrum was multiplied by this shape. A Fabry-Perot profile was then generated with the same finesse and inter-order spacing as the etalon used in our experiment. The profile was shifted to give a peak at zero wavelength shift, and then the filter-modified spectrum was convolved over a thousand points with the F-P profile, simulating the scanning operation. Finally the resulting spectrum was divided by a filter profile as in the reduction (widened approximately by the instrumental resolved width). The result is shown in Figure 3.16. To this curve was fitted a Gaussian in the usual way, also shown in the figure, and it can be seen that the correspondence is remarkably good.

It is not difficult to show that for a fairly wide range of initial absorption line shapes, and even for close double features such as might be expected if the dust is moving in both prograde and retrograde orbits, the result as seen by the instrument is still very nearly Gaussian in shape. Thus the use of fitted Gaussian curves to measure the parameters of the spectra not only provides an unbiased mathematical assessment more reliable than estimates by eye, but represents a very close approximation to the physical reality. My program, GAUSSN, for computing a least-squares fit Gaussian curve to the experimental points, is shown in Appendix 2. GAUSSN was used in each case to output the parameters arrived at in the fitting process - new and more accurate estimates of the parameters of the absorption feature for all the averaged scans. The complete set of these averaged scans is shown at the beginning of Chapter 4, plotted, each with a Gaussian curve fitted, by program GAUSPT, except in two cases where no good fit could be computed; in these cases polynomial curves have been used, as noted.

At this stage, the observations were fully reduced, and in a form suitable for comparison with theoretical models of the Zodiacal Dust Cloud.

Chapter 4
Interpretation of results in terms of physical models

In this chapter, I present the fully compensated averaged scans obtained in the two observing periods, 1971 and 1972, for all observed elongations, each with a fitted Gaussian curve, except as noted. From parameters derived from these fits, plots of average wavelength shift, line width and line depth against elongation are constructed. These observations, the Hicks, May and Reay (HMR) data, are compared with previous and all subsequent observations since 1974, and examined with reference to the predictions of various models. The models are based on various distributions of dust moving in prograde and retrograde solar orbits, and a new model is presented of interstellar material flowing through the Solar System. A geocentric component, and the effects of distant emitting matter are also considered. New theoretical models by various authors since 1974 are examined and compared with the data. Theories are advanced as to why there might be seasonal variations in the radial velocities measured in the Zodiacal Cloud (ZC).

4.1 Gallery of Spectra

The following 10 pages contain reproductions of the whole set of averaged spectra, fully compensated for filter shape, overall light levels, and daylight contamination. They show intensity plotted against wavelength for all targeted elongations, in the two observing periods, September-October 1971, and April 1972. For ease of comparison, the two sets of data are matched side by side according to elongation. For most spectra, a Gaussian curve was satisfactorily fitted by the method of least squares, as described in Chapter 3, and is shown in the relevant graph. In four cases, where the continuum levels were very different on each side, no Gaussian fit could be obtained, and for these, a polynomial fit was used instead. Some scans were performed over a wider wavelength range than 5 Å, but for consistency only the central 5 Å were used for this analysis. In any case, the extreme outer wavelengths were observed at low transmission of the broad interference filter, and, hence, being subject to large random errors, would not have improved the fits.

FULLY CORRECTED SCANS: Page 1

with fitted Gaussian curves, except where otherwise stated

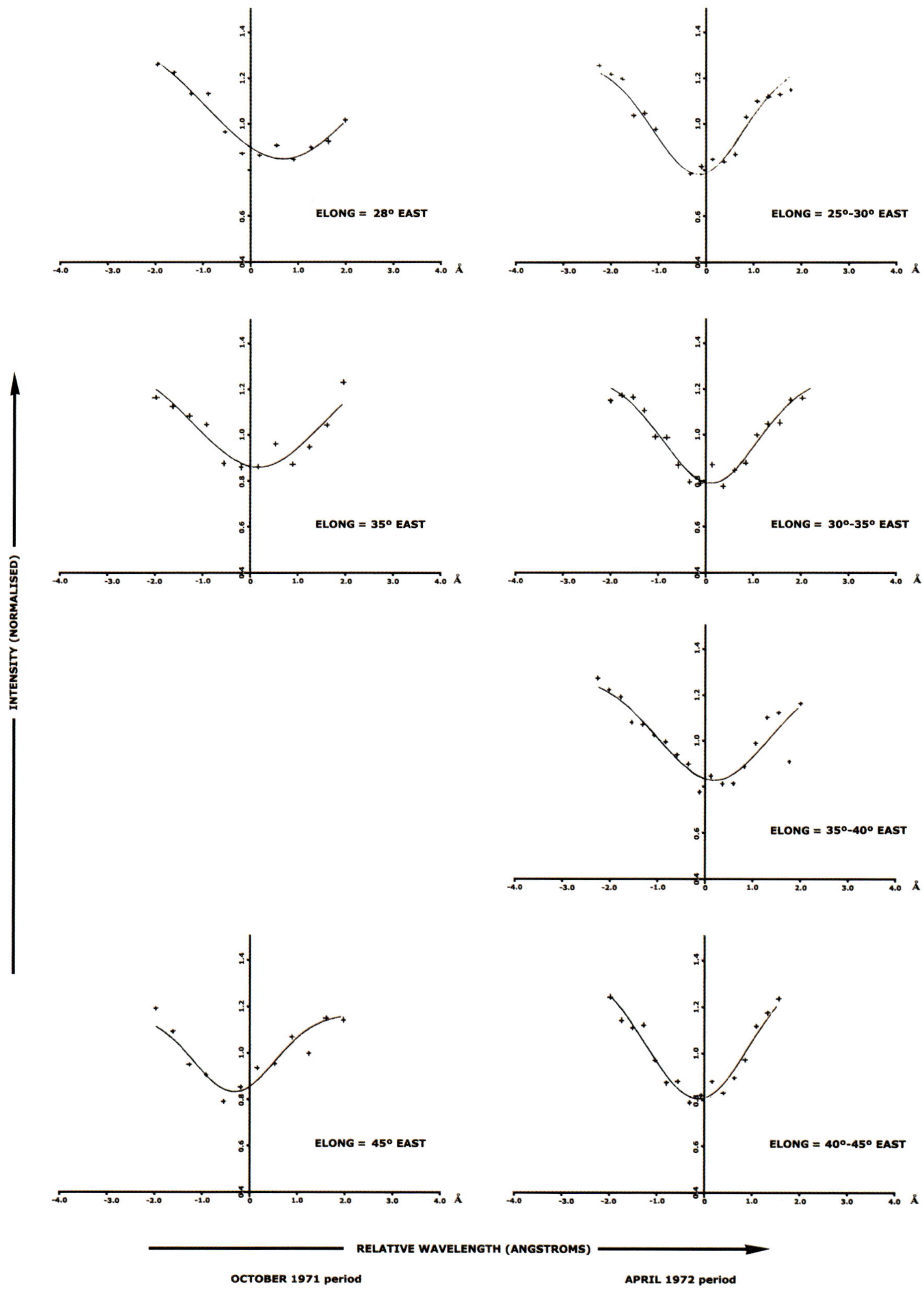

FULLY CORRECTED SCANS: Page 2

with fitted Gaussian curves, except where otherwise stated

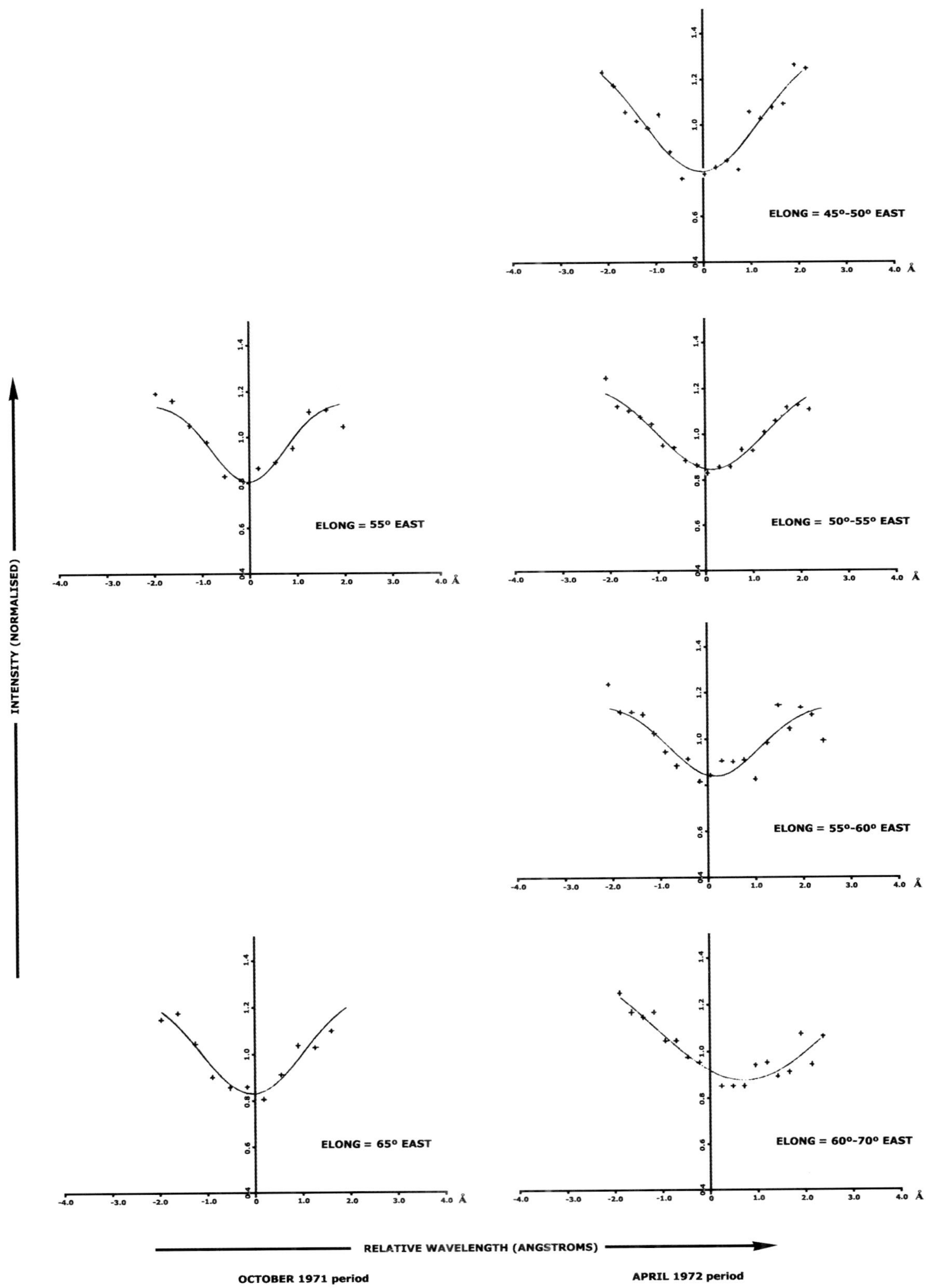

FULLY CORRECTED SCANS: Page 3

with fitted Gaussian curves, except where otherwise stated

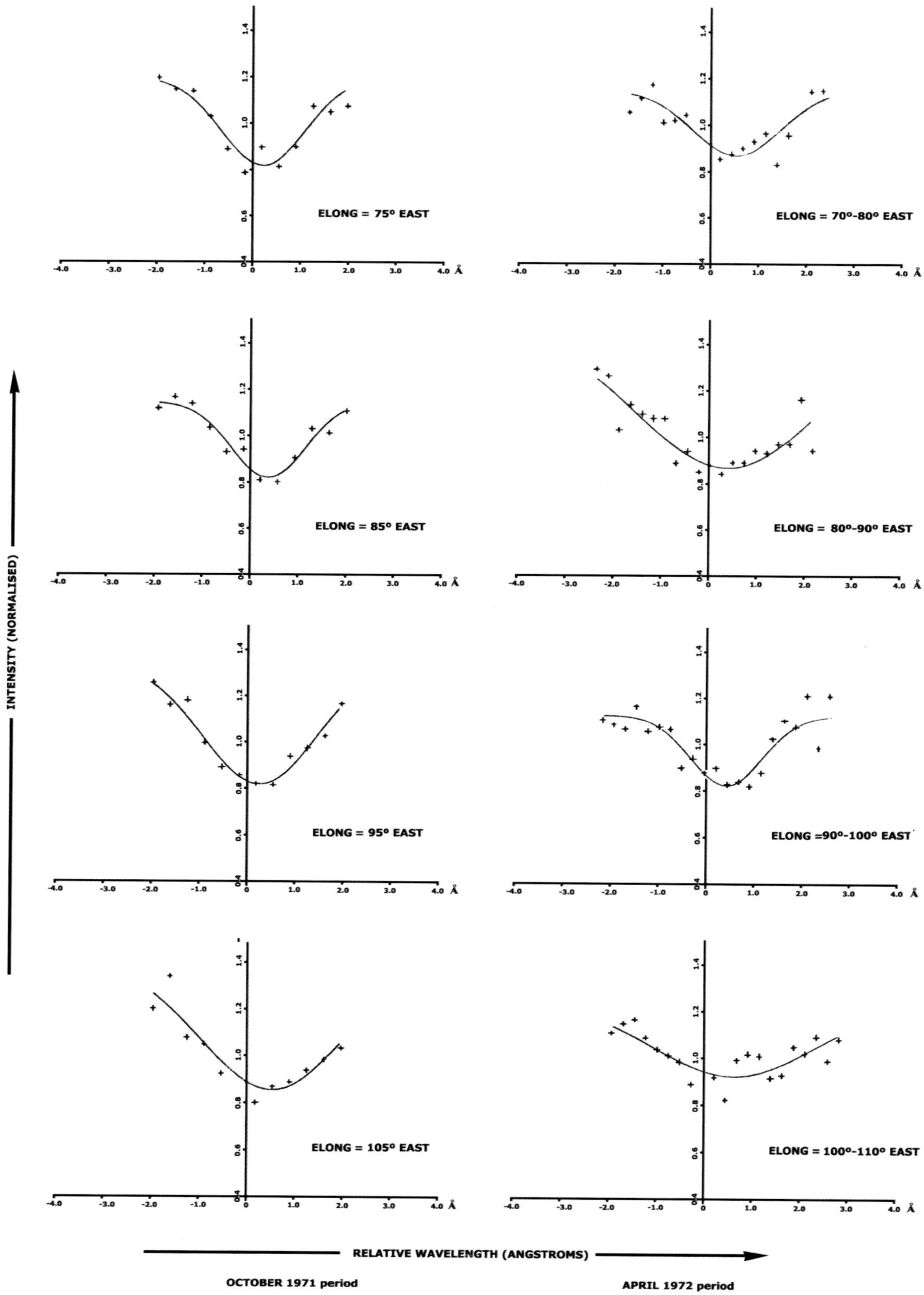

FULLY CORRECTED SCANS: Page 4

with fitted Gaussian curves, except where otherwise stated

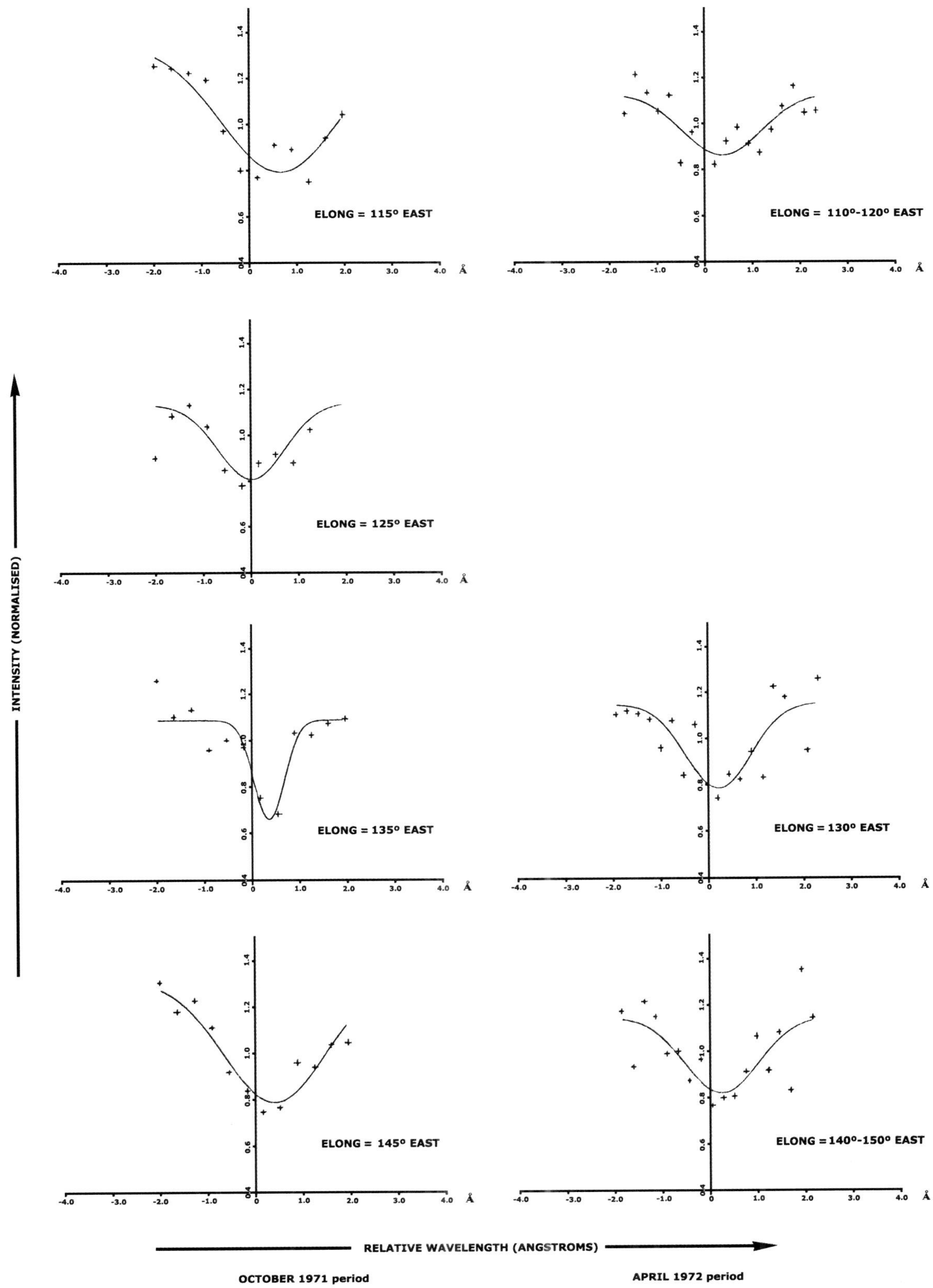

FULLY CORRECTED SCANS: Page 5

with fitted Gaussian curves, except where otherwise stated

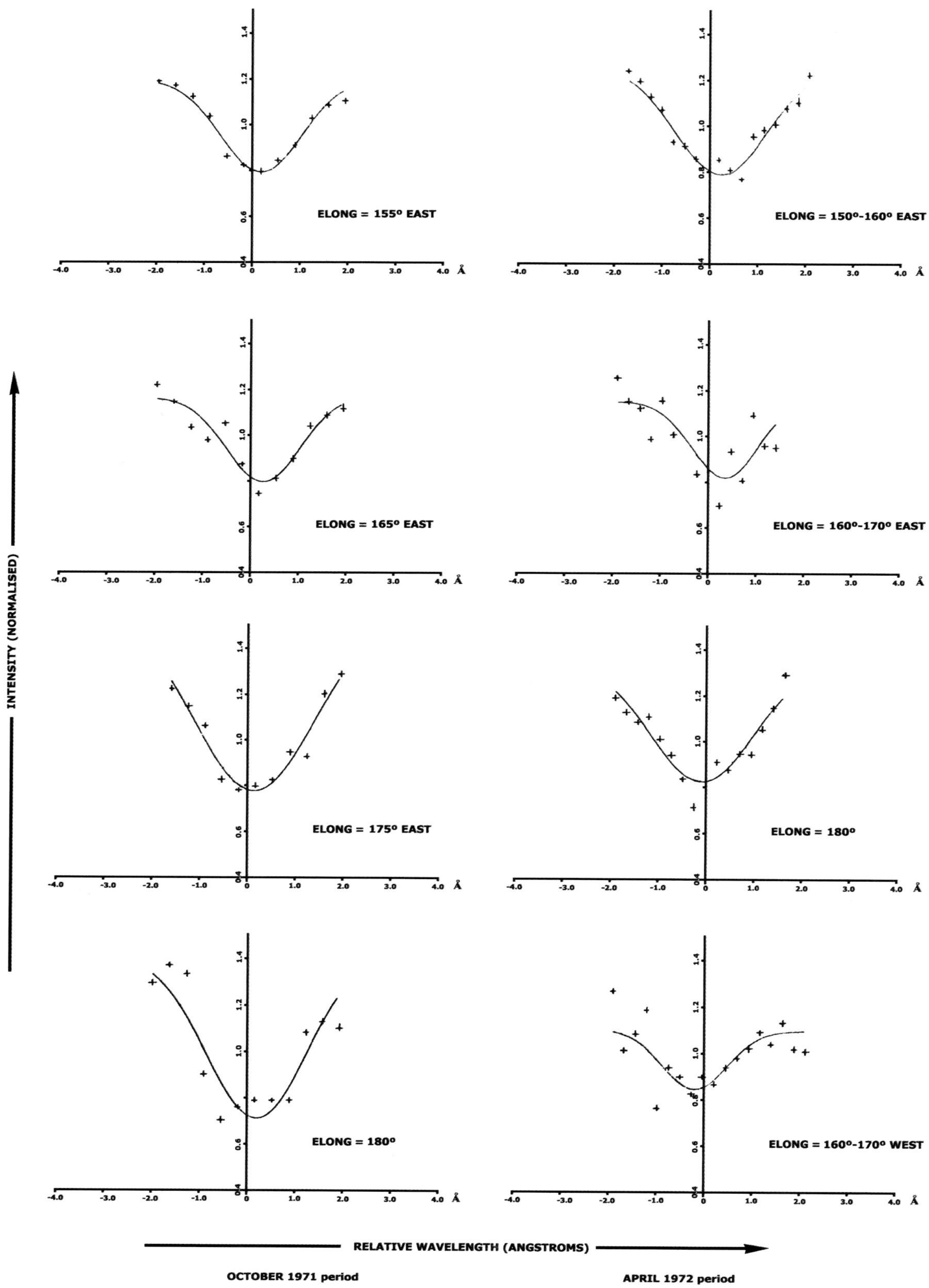

FULLY CORRECTED SCANS: Page 6

with fitted Gaussian curves, except where otherwise stated

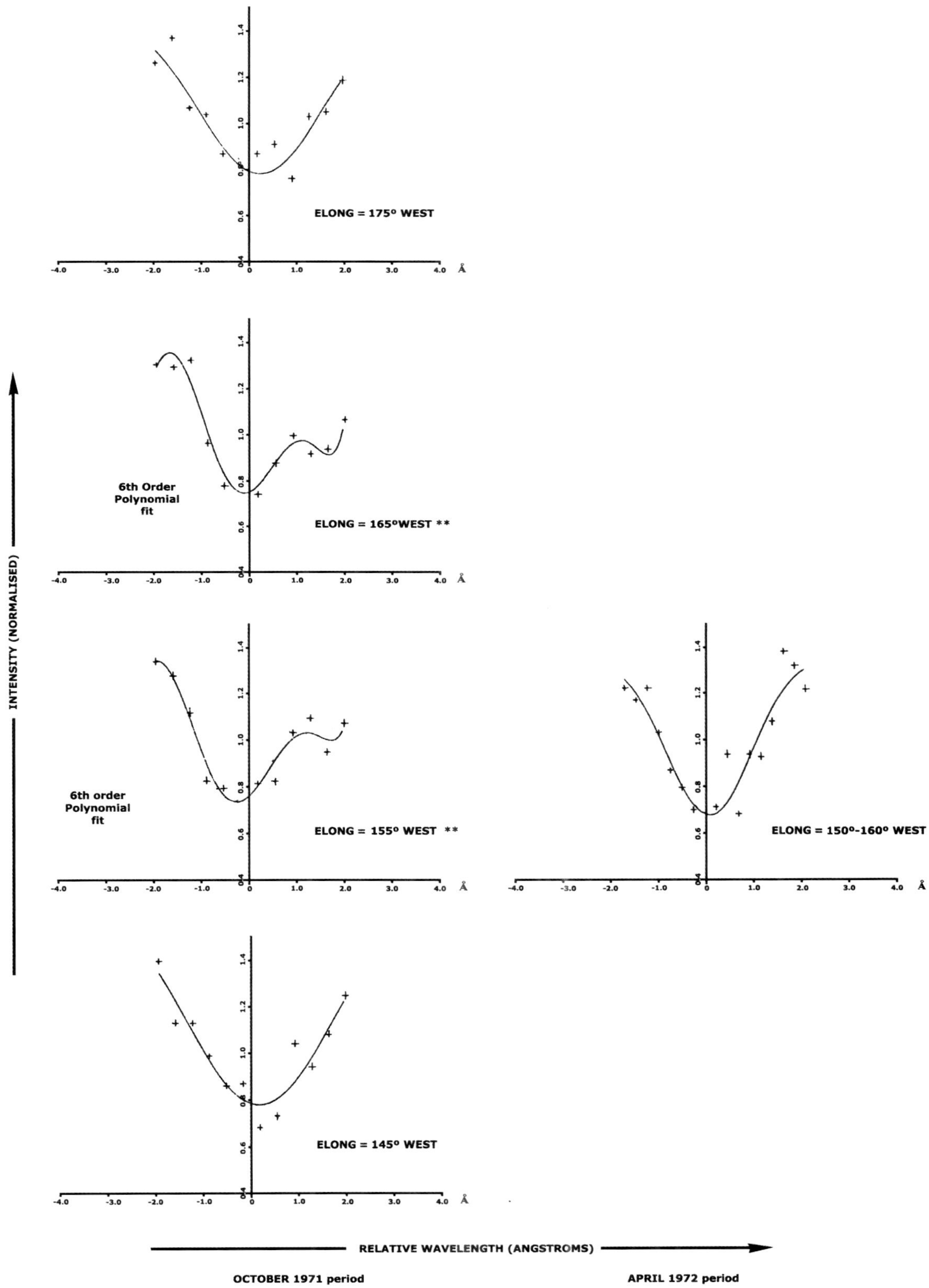

FULLY CORRECTED SCANS: Page 7

with fitted Gaussian curves, except where otherwise stated

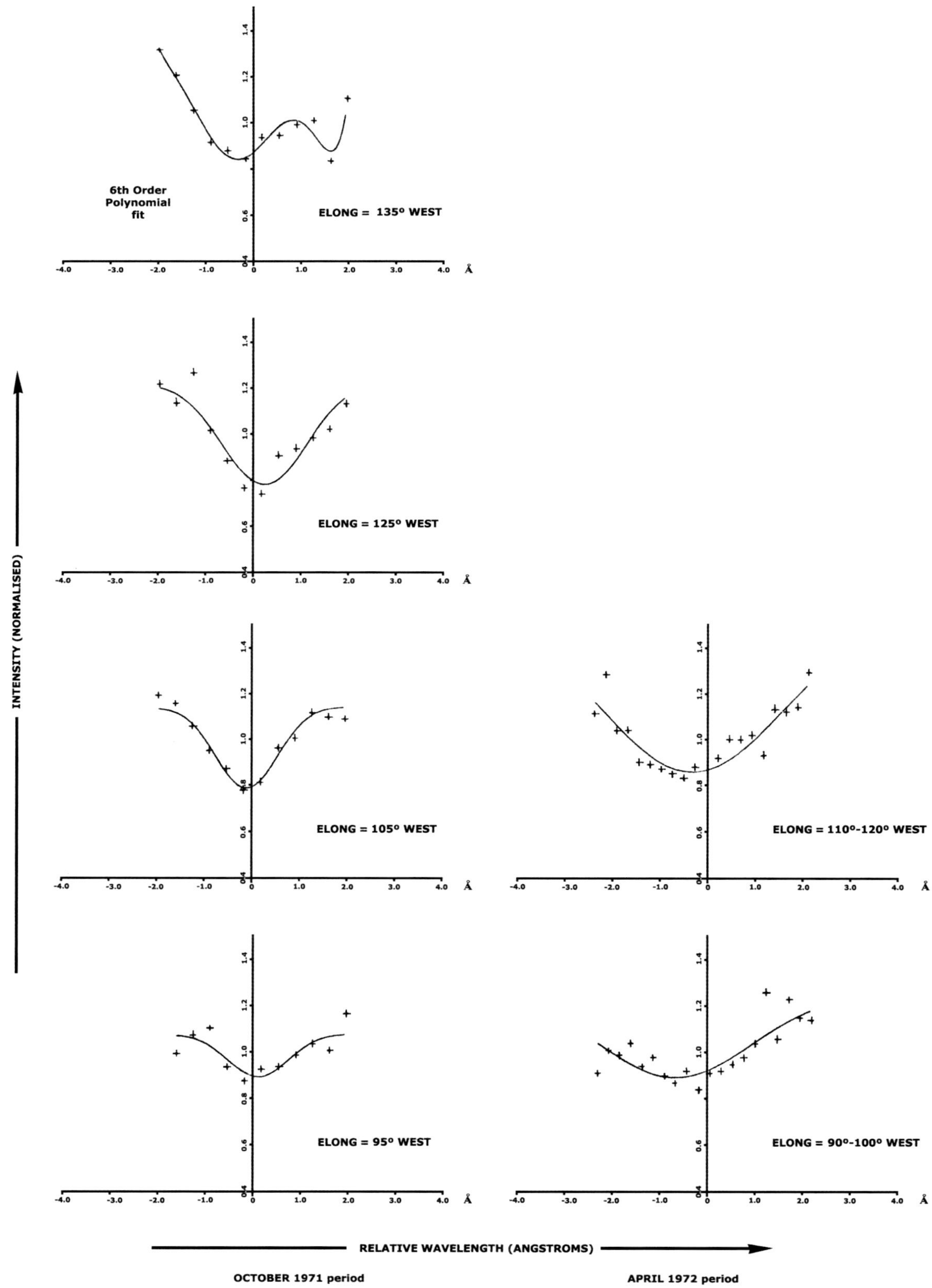

FULLY CORRECTED SCANS: Page 8

with fitted Gaussian curves, except where otherwise stated

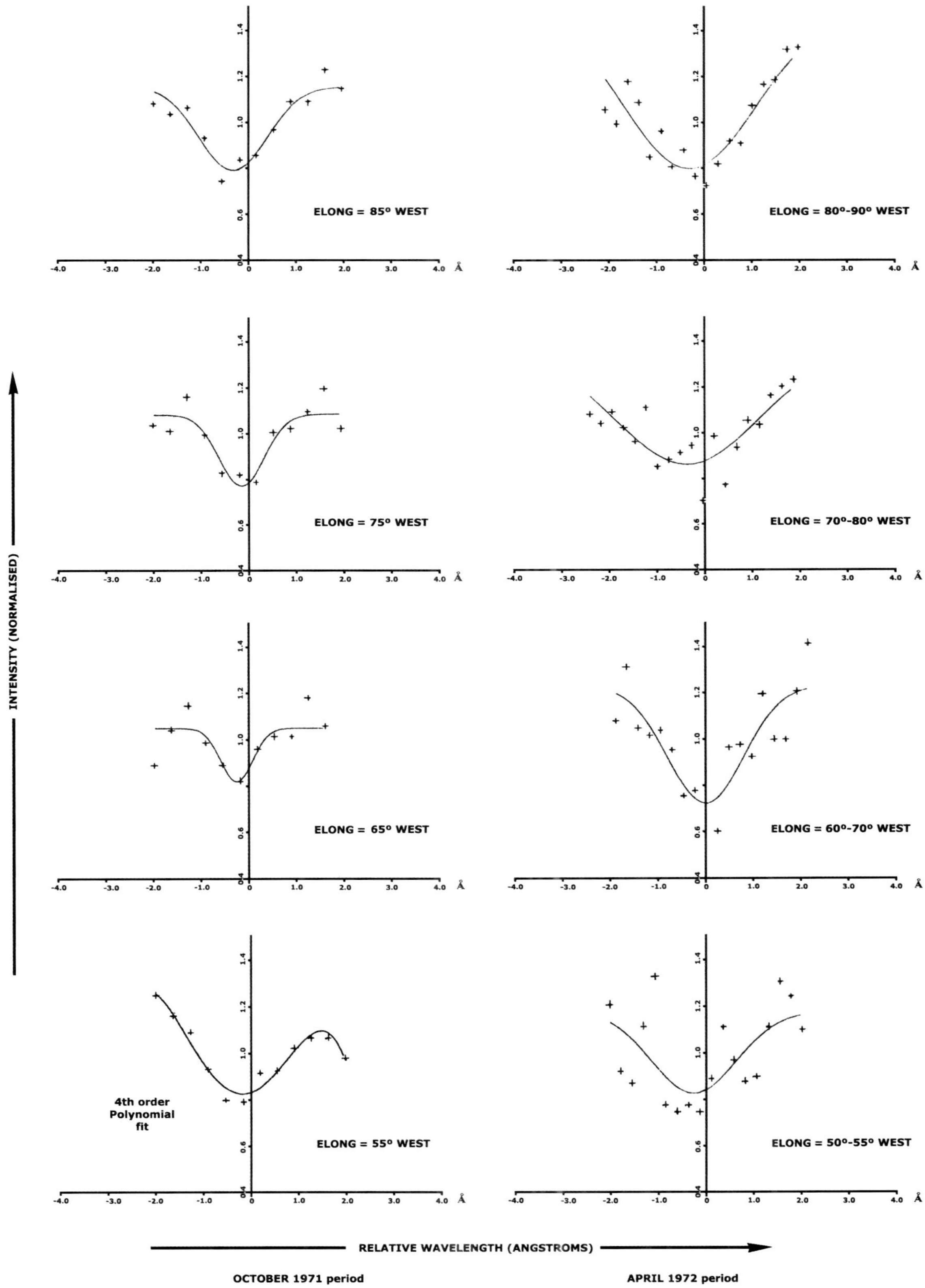

FULLY CORRECTED SCANS: Page 9

with fitted Gaussian curves, except where otherwise stated

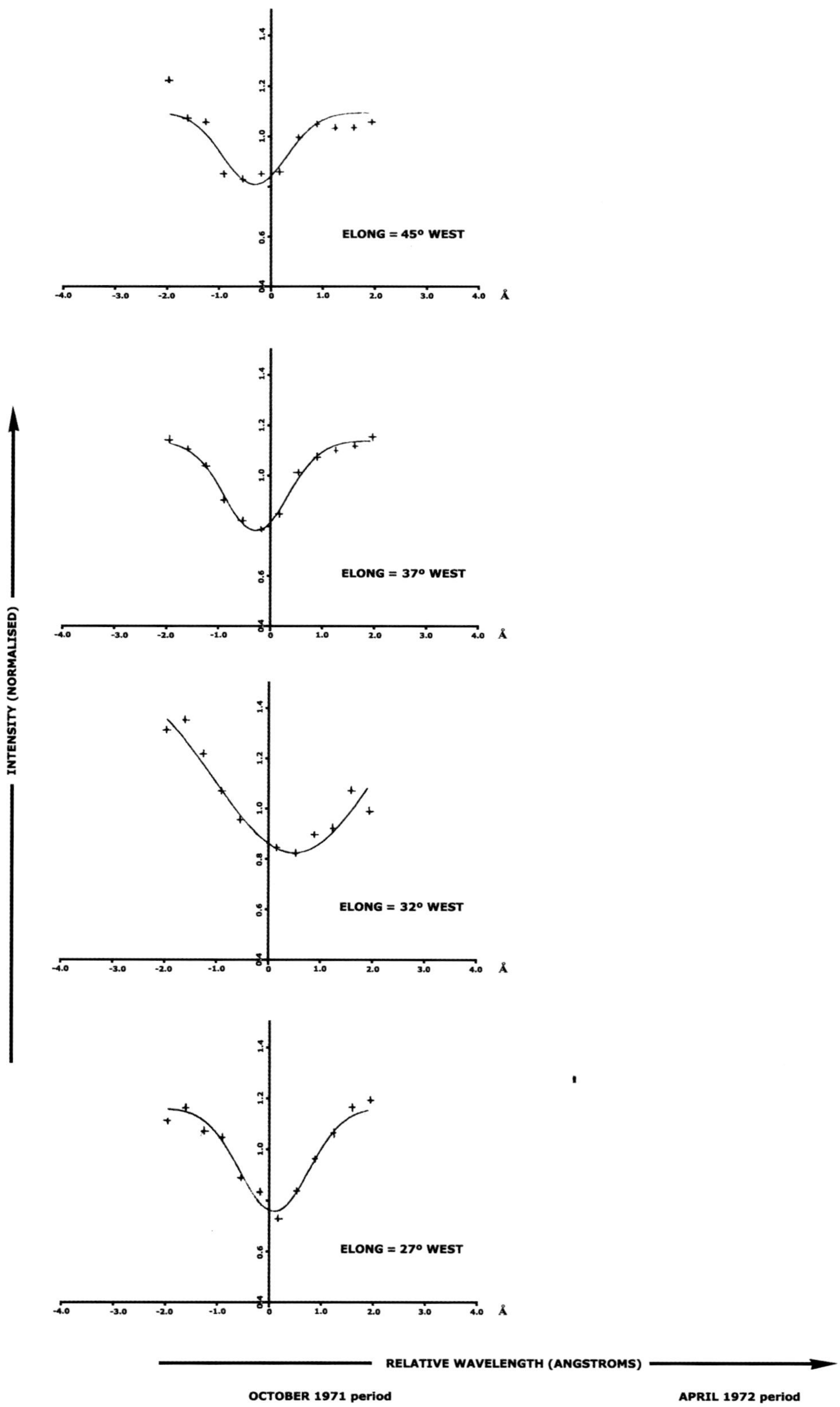

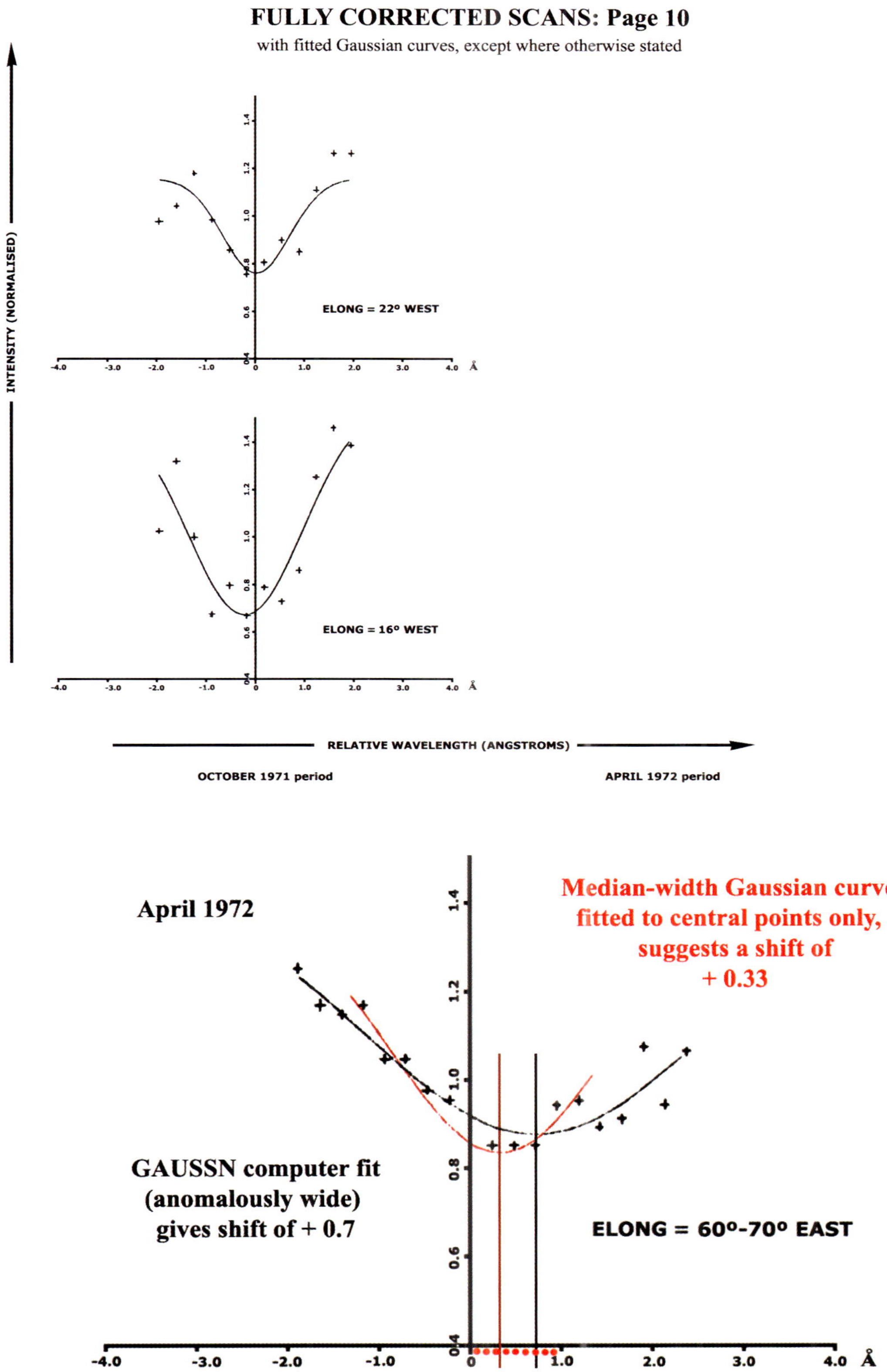

Figure 4.0 Example of an unrealistically wide fitted Gaussian curve, in this case corresponding to 'rogue' point (a) in Fig 4.1.

4.2 General comments on the data

All the averaged experimental scans show a major absorption feature in the neighbourhood of 5183.6 Å. In three cases, there is a hint of a second small absorption line, towards the red. A curious property of the data is that the scans tend to be of less overall height at large elongations. These scans have been compensated for variation in the overall level of light intensity, as monitored by the compensation channel, so this inequality is not the result of variations in intensity around the MgI wavelength. It seems likely that the 'comp' channel, with its relatively wide pass-band green filter, is picking up airglow lines, light pollution and faint stars, which, when the ZL intensity is low, significantly inflate the compensation channel counts. This reduces the apparent height of the corrected spectrum. At the lower elongations these contributions are swamped by ZL continuum, and the corrected height rises to its 'normal' value. An additional slight rise in intensity at the position of the Gegenschein is also explainable by this reasoning. In future this problem could be avoided by using a narrow filter for the comp channel, closer to the width of the pre-monochromating filter of the F-P.

The results of plotting the wavelength shifts from these fits, against elongation, are shown in Figure 4.1. There is quite a large scatter in the shift values in both observing periods, greater than that suggested by the error bars. In the case of September-October 1971, this may be the result of an unstable Fabry-Perot. However, in April 1972, the finesse stability had been significantly improved, and we might have expected a smoother curve. In fact, the points for April do follow a curve quite well, within the error bar predictions, except for a few 'rogue' points (a, b, c in Figure 4.0). Reference to the widths of the Gaussian fits in Figure 4.2 shows a very clear grouping of the widths into a cluster around the median (about 2 Å), and a few well outside this width. I think it is quite possible that this unrealistic width in a few of the fits indicates faulty curve fitting. Figure 4.0 shows such a scan. If a median-width Gaussian fit is made to the central points only (the fringe points may be even more noisy than realised at the time, as noted in section 4.4, owing to cosmic rays), in this case at least, a much smaller predicted shift seems likely. This suggests that it might be worth re-reducing these data with a prior bound set on the Gaussian width, and that this would reduce the number of outliers. For the purposes of this thesis, the 'raw' output shifts of program GAUSSN are used.

4.3 Information obtainable from the Gaussian fits

Figure 4.1, then, shows the result of plotting the mean wavelength shift $\langle \Delta\lambda \rangle$, against elongation, for the October and April periods. In view of the possibility of Zodiacal Cloud geometry and density being seasonably variable, the two sets of data are kept separate. On each diagram, the previous results of Reay and Ring (1968) and James and Smeethe (1970) are included for comparison. The broken curve drawn through the points in each case is a third order polynomial fitted to these new data. The polynomial fits were made using a slightly modified version of LSQFIT and were weighted inversely as the square of the standard error as calculated for each point by program GAUSSN. This graph (Figure 4.1), showing wavelength shift as a function of elongation, supplemented with various theoretical curves with which my data is compared, **is the basis for most of my subsequent interpretation in this study**. It is also the basis on which all other studies of radial velocities in the ZL to date may best be compared.

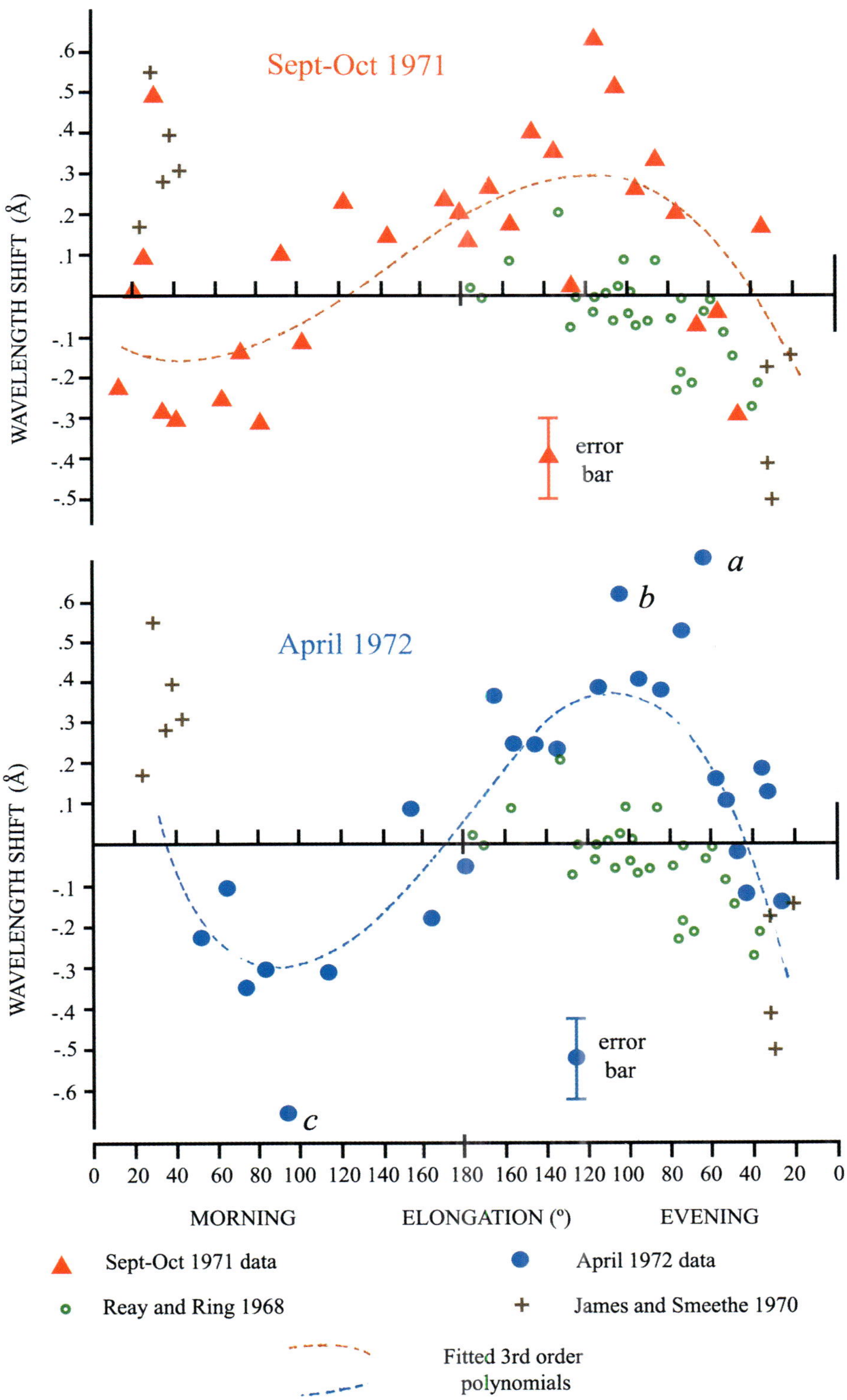

Figure 4.1 The Hicks, May and Reay data (HMR) - average wavelength shifts, determined from Gaussian fits, plotted as a function of elongation from the Sun. 'Rogue' points a, b, and c may be the result of failure in curve-fitting (see section 4.2).

Before we look at this graph in detail, two other experimental measurements, the line depth and line width of the absorption feature, can be extracted from parameters of the fitted Gaussian curve, for comparison with theory. The full width of the line at its *'half-power'* points is related to V_3 as follows:

$$I(\lambda) = I_1 - I_2 \exp\left[-k(\lambda - \lambda_m)^2\right]$$

is at a minimum when $\lambda = \lambda_m$

So ... $$I_{\min} = I_1 - I_2$$

We require the values of λ which bring the function half-way down to the value

$$I_1 - {}^{I_2}\!/_2$$

so for these points

$$e^{-k(\lambda-\lambda_m)^2} = {}^1\!/_2$$

$$\therefore \; -k(\lambda - \lambda_m)^2 = -\log_e 2$$

$$\therefore \; \lambda - \lambda_m = \sqrt{\frac{\log_e 2}{k}}$$

Now $\lambda - \lambda_m$ is just the half-width required, so

$$\text{Full width to half power} = 2\sqrt{\frac{\log_e 2}{k}}$$

$$= \sqrt{\frac{2.77260}{k}}$$

Again, referring to the Gaussian function, an exact measure of the percentage *depth* of the line (residual intensity) is evidently

$$\%\ depth = \frac{I_2}{I_1} \times 100$$

In Figure 4.2 are plotted the widths of the lines calculated from k as described above, again as a function of elongation; Figure 4.3 shows the percentage depths in the same way.

4.4 Accuracy of the fits

The program GAUSSN computes confidence limits for all four parameters, which depend on the scatter of the experimental points and on how closely the spectra approximate to a Gaussian shape. However the errors on the points are not all purely due to random shot noise. A serious source of error arises in the outside points. These are already subject to the largest shot noise, due to their small counts, but if the filter shape is only slightly misplaced in wavelength, in the dividing out process, these points are seriously affected, giving an overall asymmetry to the spectrum. To estimate the effect of this, the filter was deliberately misplaced on some test scans by a wavelength interval corresponding to the maximum drift of the etalon during a scan. To these false results the Gaussian curves were fitted as before, and the variation in parameter λ_m, and therefore of the 'average shift', was found to be about 0.2 Å maximum. This, then, is the estimate of the error bar, shown in Figure 4.1. Similar considerations apply to the error bars in Figure 4.2 and Figure 4.3. As noted in section 4.2, there is more scatter in the points than the error bars would lead one to expect. In the 1971 observations, this may be due to an F-P which was not temperature stabilised, and whose finesse was not reliable. In the April observations, with an improved F-P, I now believe that cosmic rays, not easily distinguishable from ZL signal, may have made the peripheral points in the scans even more noisy, sporadically, than previously believed, leading to some false Gaussian fits, sometimes signalled in the graphs by unrealistic widths.

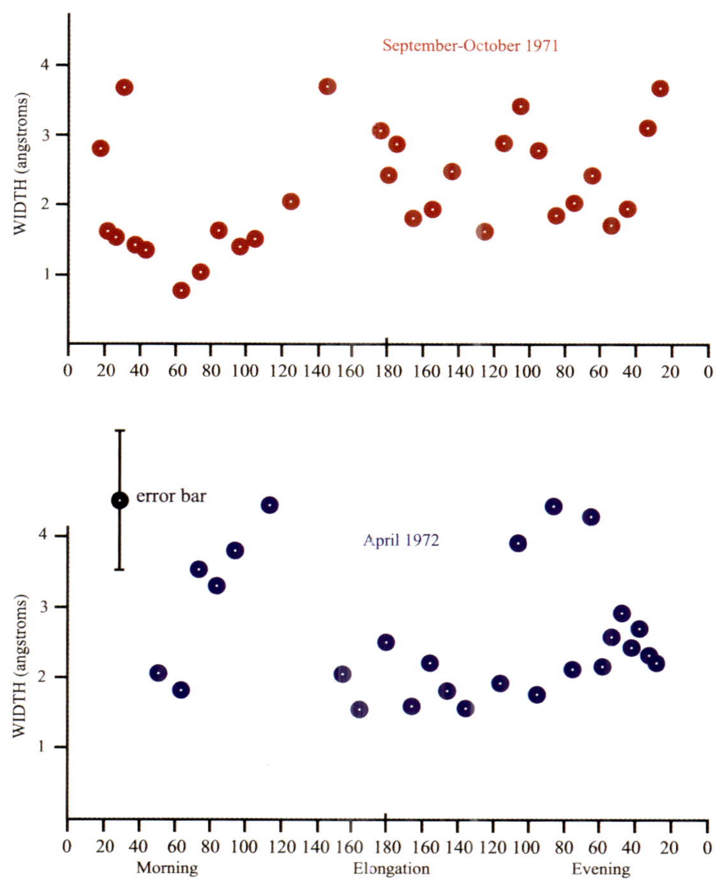

Figure 4.2 Experimental line widths computed from Gaussian fits – plotted as a function of elongation.

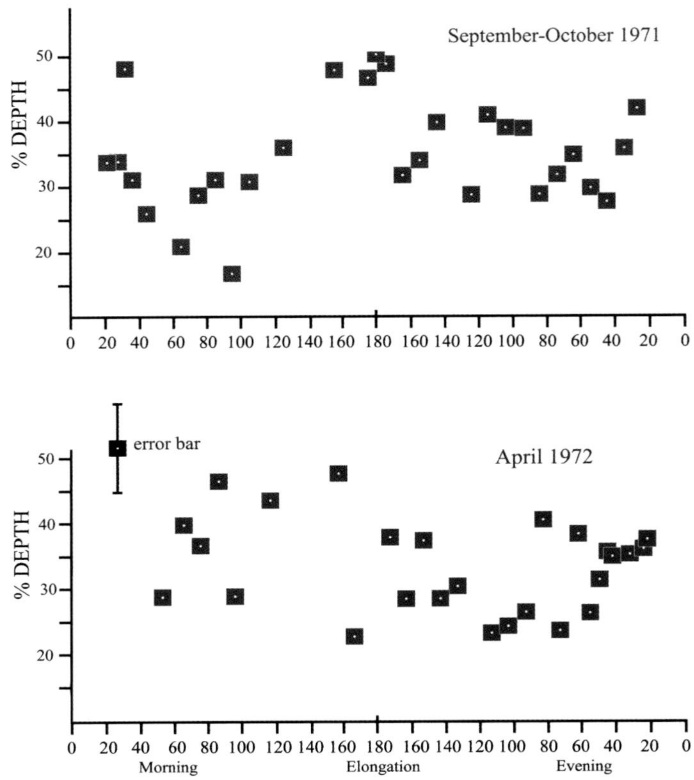

Figure 4.3 Experimental line depths of the absorption feature as a percentage of continuum level, plotted against elongation

It is worth noting that asymmetry in itself cannot be assumed to indicate a misplacement of the filter. The effect can also be the result of not sampling over a wide enough range to reach the continuum level, coupled perhaps with a chance low placing of the end points. It could also be the result of a blend in the absorption feature at the 'low' end, only partially resolved. Subsequent to my researches, Clarke *et al* (1996) predicted the possibility of real pronounced asymmetries resulting from pure prograde models with elliptical orbits and a more complete scattering theory; effectively two shifts, due to two 'cells' at opposite ends of the line of sight, are superposed. Also James (1969) was able to predict asymmetries by including retrograde as well as prograde particles. Clarke's predictions would be noticeable only at lower elongations than were observed in this study, and we will return to the question of whether the asymmetries in the HMR are significant or not, in discussing the findings of Madsen *et al* (2006), in section 4.13.3.

4.5 Evidence from line widths and depths

The predictions of Reay (1969) were used for a preliminary comparison with my data. Reay's programs were used to simulate spectra from a typical prograde circular model with $\alpha = 1$, $p = 1$, and a model similar but consisting of 50 per cent prograde material and 50 per cent retrograde. At the resolution used in the experiment, the two lines one might expect from the second model were not resolved; the result was a curve similar in shape to that due to a pure prograde model (i.e. roughly Gaussian) but slightly wider and less deep.

The predicted figures, for elongation 90 degrees, are

Width (Å)	Depth (%)	
1.86	30.4	Prograde
2.12	26.5	Prograde and Retrograde

The difference is about 10 per cent in each case. Our experimental values for these parameters are shown in Figures 4.2 and 4.3. The scatter on the points is quite large, as would be expected, and a difference of 10 per cent would not show up well in these conditions. The HMR results, as can be seen in Figure 4.2, may favour the prograde situation, but it is evident that more and better measurements are needed to make a reliable choice between models. Overall, the widths are similar to the theoretical estimates, but the relative depths tend to be slightly greater than the estimates. This is probably a reflection of the fact, already suspected from asymmetries, that the range of scanning did not satisfactorily reach the true continuum level, producing an under-estimate of its intensity, and hence a higher percentage depth. The possibility of either diffuse 'wings' to the line, or of blends, cannot be excluded, though at this point the narrowness of the pre-monochromating filter makes firm conclusions difficult. The only other observers to publish measurements of line-widths are Madsen *et al* (2006). Their results are shown for comparison in Figure 4.4. The resolution in these observations is much better (12km/s, compared with about 90km/sec), and they also conclude that no dependence of line-width on elongation can be established, but it is very interesting to note that the general pattern of points plotted is quite similar to ours (Figure 4.3). The full-width at half maximum (FWHM) expected for dust in typical asteroidal orbits (about plus and minus 20 km/sec, giving a maximum dispersion of 40 km/sec) is significantly less than the 77 km/s of their, and our, observations.

Madsen *et al* (2006) remark that to fit the observations, we need to consider a large fraction of non-asteroidal, possibly cometary particles. Future work at even higher resolutions may reveal exactly what this pattern is. Madsen *et al* (2006) comment that a flat bottom to their profiles, particularly in the Gegenschein area, where one might expect to see a pure daylight spectrum, indicates large numbers of particles in elliptic orbits. However the theoretical curve for elliptical orbits in Figure 4.4 does not fit their observations, or ours. The main conclusion of Ipatov *et al* (2006) is that asteroids alone cannot account for the WHAM observations of radial velocities, and that particles produced by comets, including high-eccentricity comets such as Comet 2P/Encke and long-period comets, are needed. The mean eccentricity of Zodiacal dust particles that best fit the WHAM observations is estimated to be about 0.5.

Evidently any inferences from these data on the extent of 'filling in' of the Fraunhofer line by the superposition of a very blurred spectrum (Ingham 1963) would also be unreliable, and in general, though these figures are included for the sake of completeness, I have not drawn any firm conclusions from them.

4.6 Analysis of the wavelength shift versus elongation data

Since the large probable errors in the estimates of line-depth and line-width make it difficult to analyse these graphs with confidence, we will turn our attentions now to the plots of wavelength shift as a

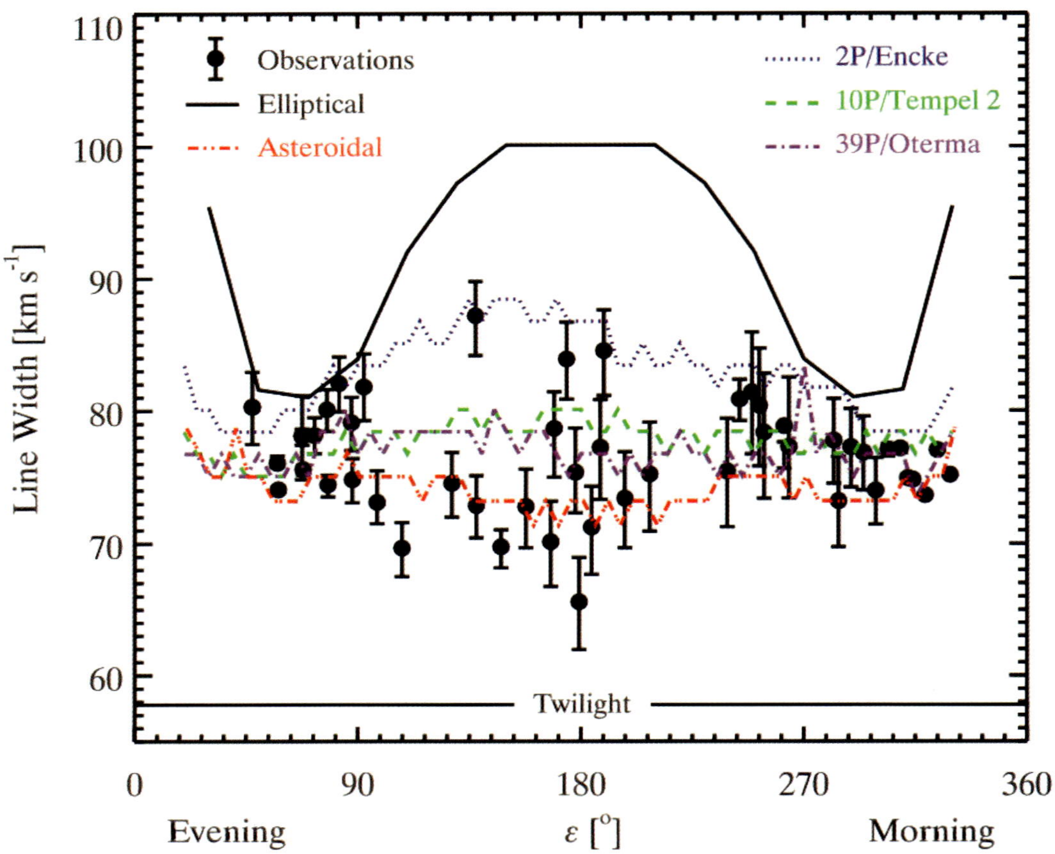

Figure 4.4 From Madsen *et al* (2006). Line widths as a function of elongation. The coloured lines are models which trace the motion of populations of dust particles of various origins, subject to gravity, radiation pressure, and drag forces, with a ratio of radiation pressure to gravitational force $\beta = 0.02$.

function of elongation (Figure 4.1), for which the analysis is much more promising. The dotted curves in this figure show a preliminary least-squares fitted polynomial, computed for each set of data, in a program adapted from the LSQFIT. Whilst the April 1972 data curve is fairly symmetric about the anti-solar point, a distinct evening-morning asymmetry appears in the September-October 1971 data. The evening radial velocity maximum at $\varepsilon = 110^\circ$ is significantly higher than the morning maximum at $\varepsilon = 40^\circ$.

Here follow various attempts to synthesize the experimental data using dynamical models of the supposed Solar System dust cloud and its environs. From each model, the behaviour of the average shift $\langle \Delta\lambda \rangle$ with elongation is predicted and compared with the shifts extracted from my data. For a more thorough comparison, a full spectrum over the range of interest could be predicted for each elongation, sampled at a rate comparable with that of the data. This kind of analysis is deferred to a future experiment.

4.7 Forces experienced by particles in the Zodiacal Cloud

In the models discussed here, the forces acting on a particle are considered to be gravity, radiation pressure, and the Poynting-Robertson effect (Robertson 1937). Forces due to electromagnetic drag, the action of the Sun's magnetic field on a charged particle, are neglected in this first analysis, which is of a quasi-steady-state velocity pattern.

This decision needs some justification. Parker (1964) calculated that U/V and soft X-rays might produce a charge of +10 Volts on a particle in orbit in the Solar System. The Sun's magnetic field, that of a dipole, but with the lines of force massively stretched out along the ecliptic plane by the Solar Wind, reverses in sign every few days for a particle in orbit, and the subsequent Lorentz force on sub-micron particles tends merely to disperse them slowly from the invariable plane.

The constant drag force has been estimated as a small force comparable with the Poynting-Robertson force by Singer (1967), and in a co-rotating plasma may be a small accelerating force.

Belton (1966), in a fuller analysis of all the charging and discharging processes, concludes that the force due to the equilibrium charge per particle at 1 AU from the Sun is of the same order of magnitude as both radiation pressure and gravity for a particle of 0.1 microns radius, although he admits to a lack of information especially with regard to discharge, which depends strongly on the shape of the particle. However, Belton has underestimated the radiation pressure at 0.1 microns by using the geometrical cross-section, so the effect of the Lorentz force is over-estimated; in any case the latter decreases sharply with particle radius, so for a particle of 0.3 microns it is proportionally nearly an order of magnitude less, even on this reckoning.

Thus the typical scattering time of these forces is long, and the forces do not significantly modify the momentary trajectory of a particle, even though they play a vital part in the evolution of the interplanetary medium.

4.8 Rotating dust cloud models

Many programs have been written to predict the Doppler effects of radial velocities on the absorption line spectrum generated by various rotating dust cloud models. Some recent simulations will be discussed in this chapter, but for an initial assessment, the predictions of Reay (1969) were used. Predictions were made from prograde and retrograde models of dust in orbit around the Sun. The orbits are assumed to be circular, and the Poynting-Robertson effect is for this first analysis ignored.

The model assumes a spatial density which falls off as $r^{-\alpha}$ at a distance r from the Sun, and a distribution of particle radii a given by:

$$n(a)da = ka^{-p}da$$

The radiation pressure exerted on a particle by sunlight is proportional to the flux at the location of the particle. The force it experiences is therefore, like the gravitational force, inversely proportional to the square of its distance from the Sun, but in the opposite direction. The effect of radiation pressure is thus to reduce the effective gravitational mass of the Sun for the dust particle, *reducing* its orbital velocity for a given distance from the Sun. But the force is also proportional to the surface area offered, i.e. to the square of its radius, whereas the gravitational force is proportional to its cube, so radiation pressure becomes more important the smaller the particle is. Hence we can expect the magnitudes of the wavelength shift in the predicted curves to vary strongly with p, and on whether the orbits are prograde or retrograde. In Reay's treatment, both reflected and diffracted components of the scattered light are evaluated, and an integral along the line of sight gives an effective wavelength shift

$$\langle \Delta\lambda \rangle = \frac{\int \Delta\lambda . dB}{\int dB}$$

For the geometry of this method, which follows that of Ingham, see section 1.2.5 and Figure 1.5; and a fuller description of the line-of-sight integral process, as applied to a continuous flow model, is given in section 4.11.

Figure 4.5 shows the results of Reay's calculations for $\alpha = 2$. In both periods of observations the agreement at low elongations (less than 70 degrees) between the new experimental points and a prograde model with a high value of p look promising. For the April results, the agreement continues to be of the right general shape at all high elongations, but for the September-October results, the model clearly predicts a curve at odds with the Hicks, May and Reay (HMR) observations, which indicate a cross-over at about 80 degrees elongation on the morning side, as opposed to the symmetrical cross-over at 180 degrees predicted in the curves.

It is noticeable that the value $p = 5$ in the prograde predictions seems to be rather too small for both sets of results. This indicates that we are dealing primarily with the very smallest particles, around 0.3 microns, assuming a density of 3 grams per cubic centimetre, which can exist in orbit around the Sun. As this size is approached from above, the repulsive force due to radiation pressure increases to the point where the effective solar attracting mass approaches zero. A particle which is thus 'just' in orbit has a low orbital speed. In other words for particles of such a size, prograde and retrograde models become more and more similar to models in which the dust is moving slowly towards or away from the Sun, or drifting *en masse* through the Solar System. A linear drift model is described separately in 4.11.

4.9 Other considerations in rotating models

This discussion of the rotating dust cloud model has so far focussed on adjusting the p parameter to fit the results. Other parameters may also produce a change in shape of the curves. The number density was assumed to be proportional to $r^{-\alpha}$, where r is the distance from the Sun, and α given the value 2, for the purpose of constructing Figure 4.5. This is a reasonable value based on measurements of intensity and polarisation as a function of elongation, made by many investigators. Bandermann and Wolstencroft (1969) show the results of calculations for other values of α (the parameter is called β in this paper). The effect of decreasing α is to raise the curve, giving higher values of $\langle \Delta\lambda \rangle$ for elongations in the range $75^\circ \leq \varepsilon \leq 180^\circ$, so a model could be envisaged which did not have a predominance of sub-micron particles ($p = 0$ *or* 1, say) and a number density staying constant with increasing distance from the Sun ($\alpha = 0$). This latter is almost certainly physically untrue, however. Although some authors have suggested that the density may increase again towards the asteroid belt, all modern obser-

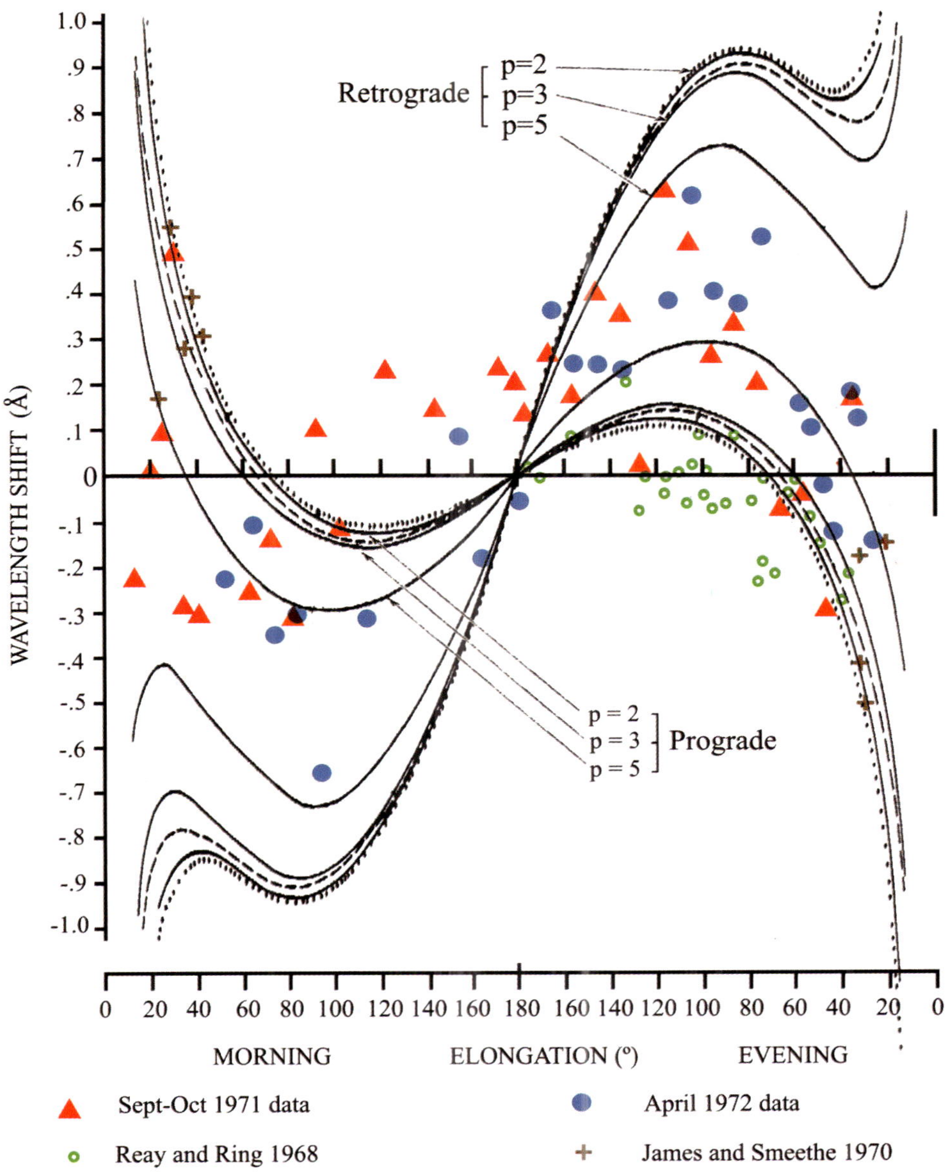

Figure 4.5 The Hicks, May and Reay data of wavelength shift vs. elongation, with Reay's predicted curves for prograde and retrograde circumsolar dust clouds, moving in elliptical orbits. Solid lines represent models with particles of size 0.2 to 100 microns; broken curves are the result of eliminating submicron particles, and dotted curves are for models in which diffraction and radiation pressure are ignored.

vations, including ISO (Reach *et al* 2003), point to a number density *decreasing* with radius from the Sun, as far as, and some way beyond, the Earth. The difference between the curves for $\alpha = 1, 2, or\ 3$ is not very great, and increasing α beyond this point makes almost no difference. The region close to the Sun, $\varepsilon \leq 70^{\circ}$, is not sensitive to radial distribution and it is in this region that the rotating model fits most satisfactorily, so the conclusion that small particles, a micron or less in radius, dominate the cloud is still valid.

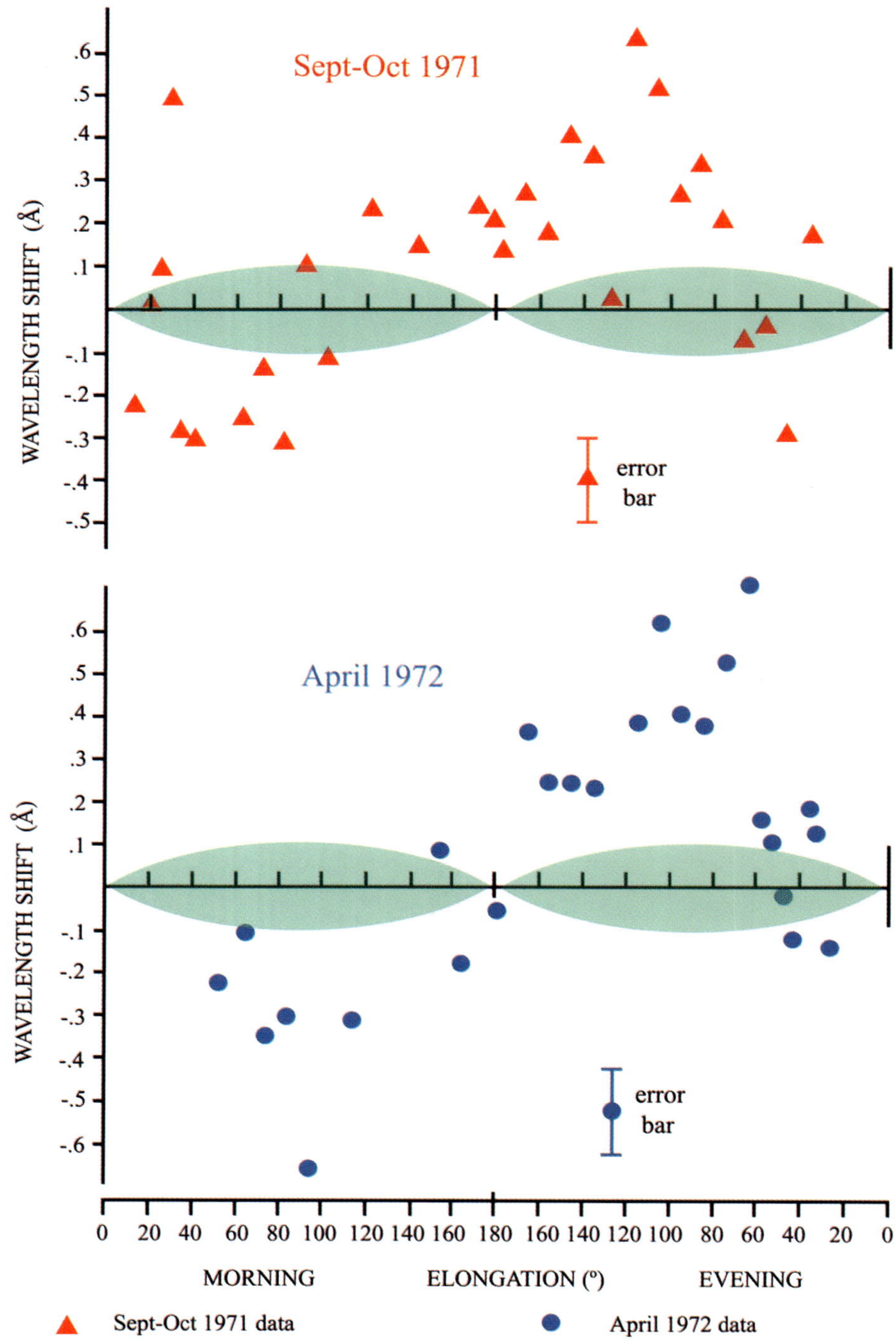

Figure 4.6 Predictions of Vanysek and Harwit compared with both sets of HMR data. The shaded areas indicate the allowed range of wavelength shifts assuming a geocentric dust cloud.

Bandermann and Wolstencroft also give predictions for elliptical rather than circular orbits. Even for eccentricities near to 1, their curves are not changed as much as for variations in α, and such high eccentricities are highly physically unlikely. And elliptical orbits in any case degenerate to circular under the

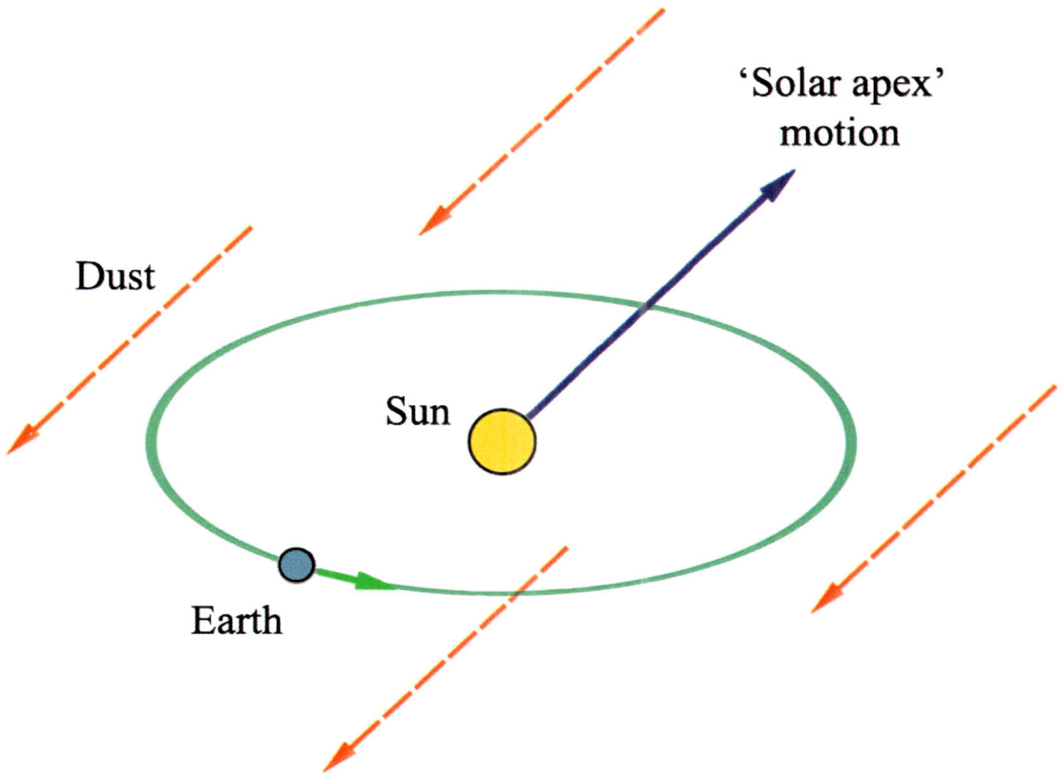

Figure 4.7 Illustrating the movement of the Sun and the Earth through the local interstellar medium.

influence of the Poynting-Robertson effect. Thus the possibility of elliptical orbits does not materially affect the validity of the conclusions already drawn here. However, we will see that Ipatov *et al* in 2006 came to different conclusions, based on a model in which the dust is being continuously replaced, and there will be much more to say about rotating models in sections 4.13 and 4.14. Meanwhile let us look at other possible models, which include no dust in orbit around the Sun at all.

4.10 Geocentric Dust Cloud (GDC) model

A thorough theoretical investigation by Shapiro, Lautman, and Colombo (1967) asserted the absence of a mechanism by which the Earth could acquire a dust cloud of the proportions necessary to make an appreciable contribution to the Zodiacal Light. In spite of this, and the discrediting by Nilsson (1966) of the satellite measurements which gave support to the speculation, the GDC theory still had vocal advocates in 1970.

Vanysek and Harwit (1970) suggested that Reay and Ring's results for $\varepsilon \leq 70^\circ$ represent the dominance of dust near the Earth. The predictions for a model based on terrestrial dust only are shown in Figure 4.6. The HMR results certainly seem to rule out the hypothesis, as can be seen from the high shifts in the observations in the region $75^\circ \leq \varepsilon \leq 120^\circ$, both in morning and evening.

It is worth commenting that Divari's model (1965) of a cloud of terrestrial dust particles in highly elliptical prograde orbits with the major axes in line with the Sun could produce a blue shift in the morning and red in the evening; but it certainly could not account for the reversal in sense of the shift at low elongations which the consensus of radial velocity measurements clearly shows to exist. It is my opinion, based on the above, that cis-terrestrial dust is not a major contributor to the Zodiacal Light.

4.11 A uniform dust flow model

It was my feeling in 1974 that a component of the ZC might well be interstellar dust, drifting through the Solar System, due to the Sun's proper motion through the local interstellar medium. I believe that this was the first time such a theory was proposed, previous discussions having been centred on asteroidal and cometal origins of the dust. It was my intention to publish this view in a further paper in MNRAS, but unfortunately my work at Imperial College ceased before this could be achieved. I was recently, however, pleased to find vindication in the *Ulysses* and *Galileo* satellite dust collection experiments conducted 20 years later, by Baguhl *et al* (1995), in which they concluded that there was evidence in the hyperbolic speed, and the distribution of directions in the impacts on their detectors, that *interstellar dust was flowing into the Solar System*. Moreover Grün *et al* (1997) were able to estimate a proportion of 30 per cent interstellar dust at 1.3 AU, and even more at greater solar distances. This section details my simple 1973 model to simulate of the appearance of an absorption line in sunlight reflected from dust flowing uniformly through the Solar System.

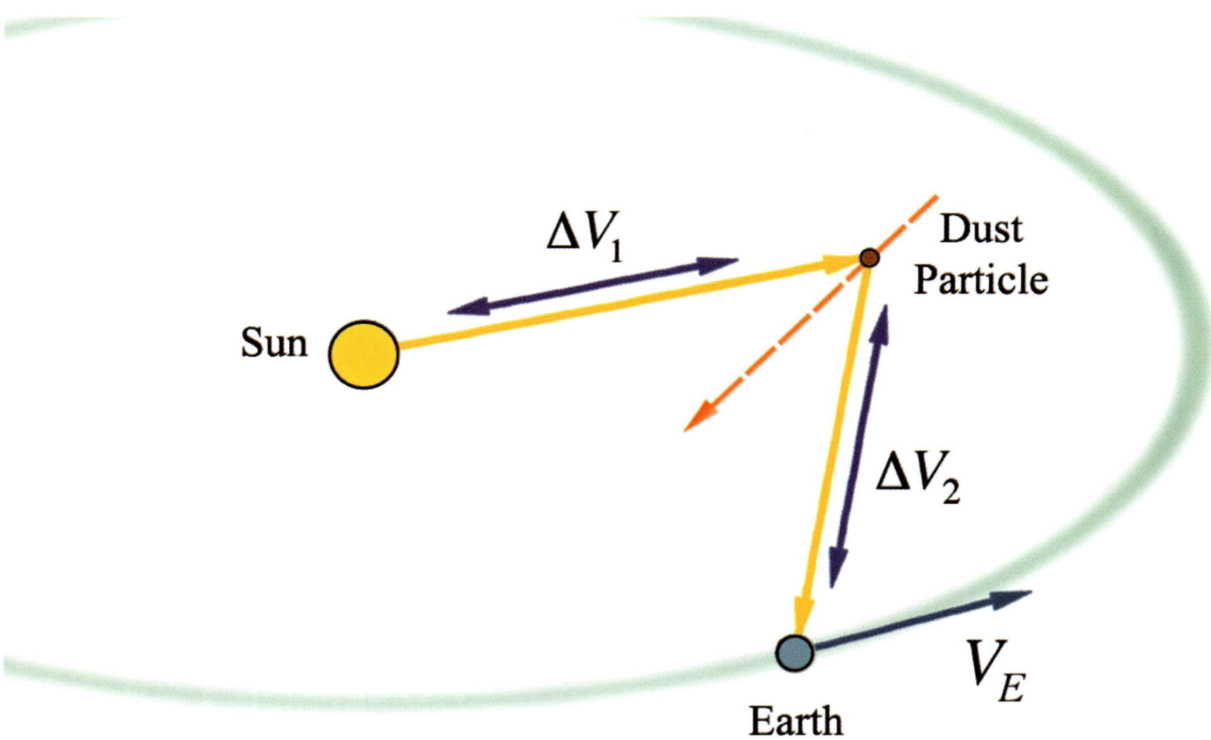

Figure 4.8 Showing the two Doppler shifts experienced by light emitted by the Sun, scattered by the dust particle, and observed at the Earth.

4.11.1 Uniform flow theory

In this analysis:

a) The direction of motion of the dust relative to the Sun is assumed to be the negative of the Sun's motion towards the Solar Apex. This is equivalent to the assumption that the dust cloud is stationary with respect to the local community of stars, and the Solar System is moving through it.

b) Focusing effects due to gravity, and scattering by radiation pressure are ignored; the dust is assumed to be of uniform density and flowing linearly.

c) Scattering is assumed to be isotropic - i.e. diffraction effects are ignored.

The method is to find the shift in wavelength of sunlight reflected from a general element in space, and integrate along a line of sight for various elongations. Only the ecliptic plane is considered.

Figure 4.8 illustrates the double Doppler shift which occurs in this configuration. Referring to the figure, light from the Sun is scattered by the dust particle, and observed at the Earth. On the way, the relative radial velocities of Sun and particle, ΔV_1, and Particle and Earth, ΔV_2 both produce Doppler shifts, $\Delta\lambda_1$ and $\Delta\lambda_2$. Considering only components of velocity in the ecliptic plane, radiation emitted by the Sun at wavelength λ is 'seen' by the particle in its own frame of reference at a wavelength $\lambda + \Delta\lambda_1$, where $\Delta\lambda_1$ is given by

$$\frac{\Delta\lambda_1}{\lambda} = \frac{\Delta V_1}{c} .$$

where c is the speed of light.

Thus $$\Delta\lambda_1 = \frac{\Delta V_1 .\lambda}{c} .$$

An observer on the Earth sees a further shift $\Delta\lambda_2$ given by

$$\frac{\Delta\lambda_2}{\lambda + \Delta\lambda_1} = \frac{\Delta V_2}{c} .$$

Since $\Delta\lambda_1$ is small, we can write λ for $\lambda + \Delta\lambda$, so

$$\Delta\lambda_2 \approx \frac{\lambda .\Delta V_2}{c} .$$

Hence the total shift in wavelength is

$$\Delta\lambda = \Delta\lambda_1 + \Delta\lambda_2 = \frac{\lambda}{c}(\Delta V_1 + \Delta V_2).$$

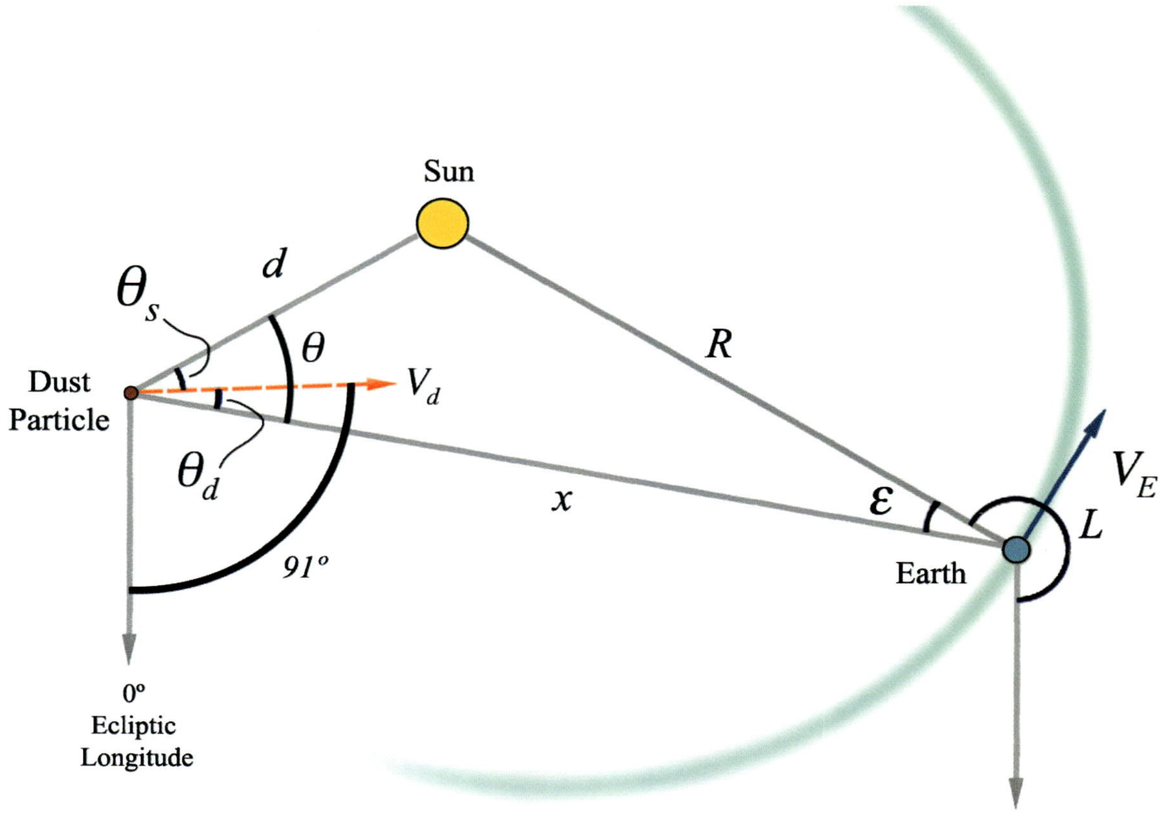

Figure 4.9 Showing the geometry of the scattering, in the plane of the ecliptic only. The diagram shows roughly an October configuration.

We require this as a function of elongation ε from the Sun in terms of the known ephemeris positions and velocities.

In Figure 4.9, which is based on the positions for the October 1st configuration, V_d is the (apparent) component of the dust particle's velocity in the Ecliptic plane. Its direction is 91° ecliptic longitude. Since all speeds are small relative to the speed of light, the effect of Special Relativity on the angles can be ignored.

$$\Delta\lambda_1 = -\frac{\lambda}{c} V_d \cos\theta_s \quad \text{and}$$

$$\Delta\lambda_2 = -\frac{\lambda}{c} V_E \sin\varepsilon - \frac{\lambda}{c} V_d \cos\theta_d$$

$$\therefore \ \Delta\lambda = \frac{\lambda}{c}(-V_E \sin\varepsilon - V_d \cos\theta_d - V_d \cos\theta_s) \ .$$

If the solar apex has ecliptic latitude B, and the speed with which the Sun is moving through the interstellar medium is S, the apparent velocity of the dust V_d is $-S\cos B$, representing the negative of the Sun's velocity component in this plane. It has a numerical value of 11.88 km/s.

It remains to find $\cos\theta_d$, and $\cos\theta_s$ in terms of ε, the elongation of the dust particle as viewed from the Earth, and x, the distance of the particle from the Earth.

If L is the ecliptic longitude of the Sun as seen from the Earth at the time in question, and since the direction of the dust flow has ecliptic longitude 91 degrees:

$$\theta_d = 91 - (L + \varepsilon - 180)$$

$$\therefore\ \theta_d = 271 - L - \varepsilon.$$

To obtain θ_s, referring again to Figure 4.9,

using the sine rule $$\frac{d}{\sin\varepsilon} = \frac{R}{\sin\theta}$$

giving $$\sin\theta = \frac{R\sin\varepsilon}{d}$$

and, by dropping a perpendicular from the Sun to the line joining the Earth and the particle,

$$\cos\theta = \frac{x - R\cos\varepsilon}{d}$$

This gives enough information to define θ for all configurations, and θ_s is then given by $\theta - \theta_d$ (the program STREAM used the ARCTAN function with the signs determined by the two functions above). By this method the Doppler shift for the particle was determined.

To find the contribution to the line-of-sight integration from an element containing such particles, referring to Figure 4.10 (overleaf), the illumination of dust in the element is inversely proportional to d^2, and if we call the scattering area per unit volume ρ, the light reflected from the element is proportional to

$$\frac{\rho dx \times x^2 d\Omega}{d^2}.$$

The illumination at the Earth due to this element is inversely proportional to x^2, so the eventual light intensity per unit solid angle of view is proportional to

$$\frac{\rho dx}{d^2}.$$

So the intensity dE due to this element, arbitrarily scaled, is given by

$$dE \propto \frac{\rho dx}{x^2 + R^2 - 2Rx\cos\varepsilon}.$$

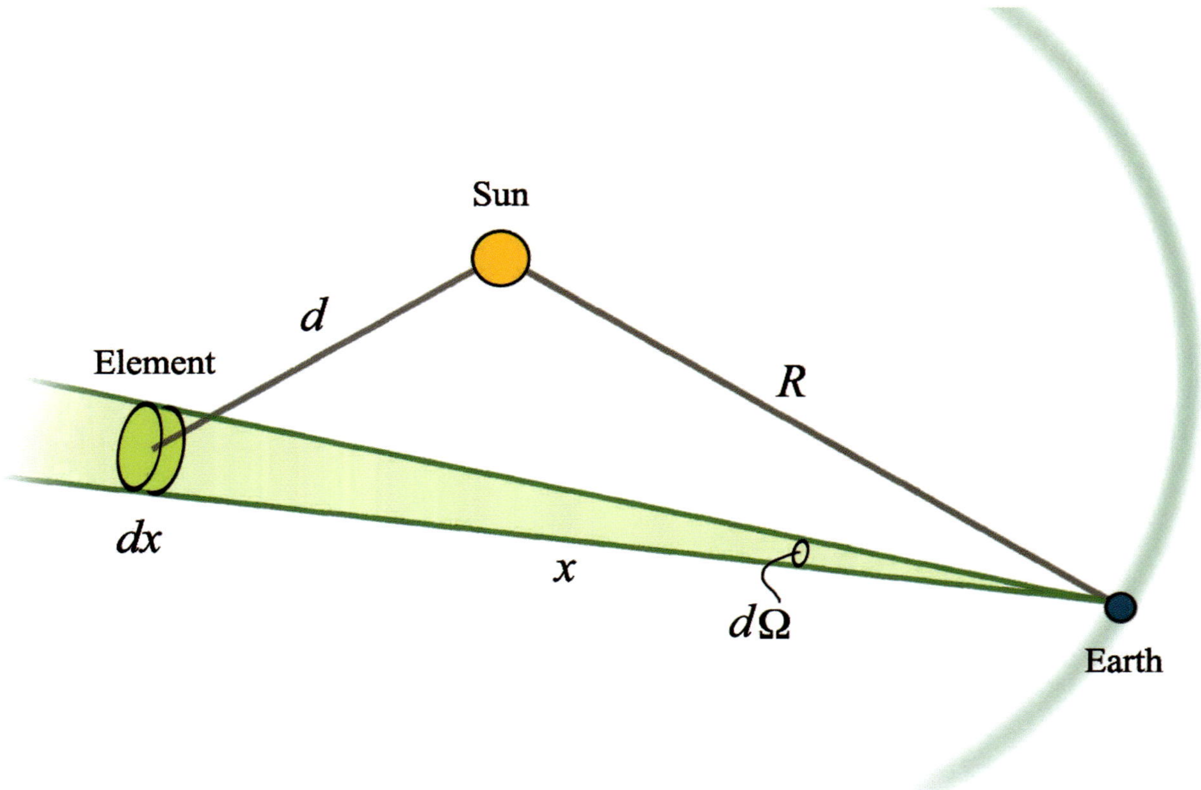

Figure 4.10 Showing the geometry of the scattering, in the plane of the ecliptic only. The diagram shows roughly an October configuration.

We can now determine the 'average shift" in the usual way as

$$\langle\Delta\lambda\rangle = \frac{\int \Delta\lambda . dE_{\Delta\lambda}}{\int dE_{\Delta\lambda}}$$

Thus
$$\langle\Delta\lambda\rangle(\varepsilon) = \frac{\displaystyle\int_{x=0}^{\infty} \frac{\Delta\lambda}{x^2 + R^2 - 2Rx\cos\varepsilon}dx}{\displaystyle\int_{x=0}^{\infty} \frac{1}{x^2 + R^2 - 2Rx\cos\varepsilon}dx}$$

where $\Delta\lambda$ is determined as above.

Program STREAM (Appendix 4) uses this method to make predictions for elongations at 10 degree intervals. The integral is approximated by dividing x into 1/10 AU steps and adding up the trapezia, the series being truncated when contributions become less than one per cent of the existing total. The output of STREAM was then processed by program CONVL (Appendix 5), which convolved the predicted spectrum with the instrumental profile of the F-P.

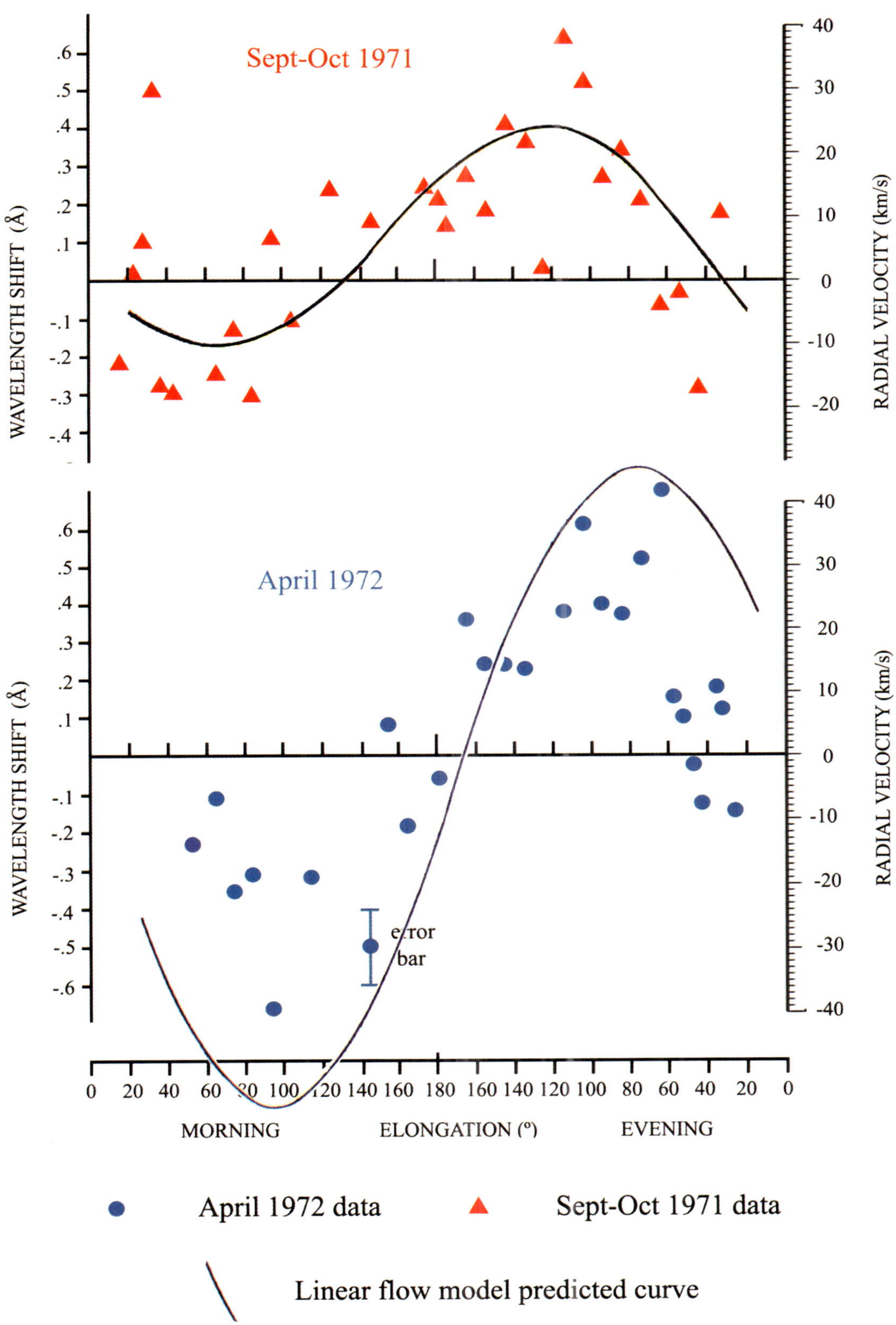

Figure 4.11 Showing radial velocity predictions as a function of elongation from the Sun for a linear dust flow model, in Spring and Autumn configurations.

The predicted curves are shown in Figure 4.11 (previous page), for 1st October and 1st April, along with the HMR experimental points for each case.

The region around $\varepsilon = 0$ exhibits one of the relatively few examples in Physics of a finite discontinuity. This is to be expected because at low elongations only dust near the Sun needs to be considered, this dust being so highly illuminated by the Sun, as dictated by the inverse square law of illumination. The effect is, however, not a point discontinuity, because of course the Sun appears not as a point, but with a finite angular diameter.

For a particle orbiting between us and the Sun, the Doppler shift due to the particle's motion with respect to the Sun is in opposite senses on opposite sides of the Sun, as viewed from the Earth, and as we look at smaller and smaller orbits, the Sun begins to mark a sharp, discontinuous change in Doppler shift.

In fact, the low elongation region is omitted from the diagram for a different reason. If the particles in this region are small with respect to the wavelength of light, as we have reason to believe they are, diffraction becomes important. Diffraction becomes an important consideration in general for any particles with small scattering angles – which occur closest to the Earth along a given line of sight. The half-width of the diffraction peak is given in *radians* by:

$$\frac{1.22\lambda}{d}$$ .... where d is the diameter of the particle.

For a 0.3 μ particle and for $\lambda = 0.5\mu$, this gives

$$\frac{1.22 \times 0.5}{0.3} \approx 2$$ (radians), or about 100 degrees.

The light diffracted in this case is thus still scattered over a wide range of angles, though somewhat less than the 360 degrees assumed for isotropic scattering. This, coupled with the fact that the statistical weighting for these particles is small, means that this roughly zero-shifted component does not seriously compete with the reflected components in the integral along a line of sight in the regions we have been able to observe. Very close to zero elongations this component would eventually become important enough to mask the discontinuity around the Sun, by bringing the curves back towards zero, the effect being more marked if the particles were extremely small. Perhaps, when spectroscopy of the ZL is performed from a space vehicle, it will be appropriate to return to this discussion. In the meantime there is no need to pay any further attention to diffraction in any of these models. As a final note on the situation close to the Sun, there is also reason to believe (Over 1958) that at less than a few solar radii the particles will be vaporised – another reason to ignore this region for these calculations.

4.11.2 Comparison of continuous flow model with experiment

The model gives a good fit to the experimental values for the September-October 1971 period for the high elongations where the rotating model failed to provide a satisfactory fit. At low elongations in the morning the rotating model gives a better agreement, and at low evening elongations there is little to choose between the two models anyway. The April 1972 data show some signs of higher shifts at intermediate elongations which cannot be accounted for by the rotating model, and these could be explained if *some* of the dust were in uniform linear flow; it would make sense that in these elongations we are looking at a much reduced density of orbiting particles, as assumed in all the models, and a constant density of interstellar immigrants, giving them a relatively higher percentage of the total in this region. Again, the rotating model satisfactorily predicts the low elongations. The assumption of the presence of a greater percentage of flowing dust in September-October 1971 than in April 1972 could account for the lack of consistency in fit to the rotating dust model. The Autumn continuous flow curve in Figure 4.11 has another interesting feature: apart from giving a creditable visual fit to the HMR data, it gives the right kind of morning-evening asymmetry, and a crossover well into the morning side, giving a recessional speed for the Gegenschein. These features were both to be confirmed later by East and Reay (1984), as will be discussed in section 4.13.2.

It seems, then, that a *combination* of the rotating and continuous flow models might give the best fit to these data. A ten per cent contribution of continuously flowing dust produces an effect similar to that seen in our results, but higher quality data, over a longer period, would be needed, to put any certainty into this estimate.

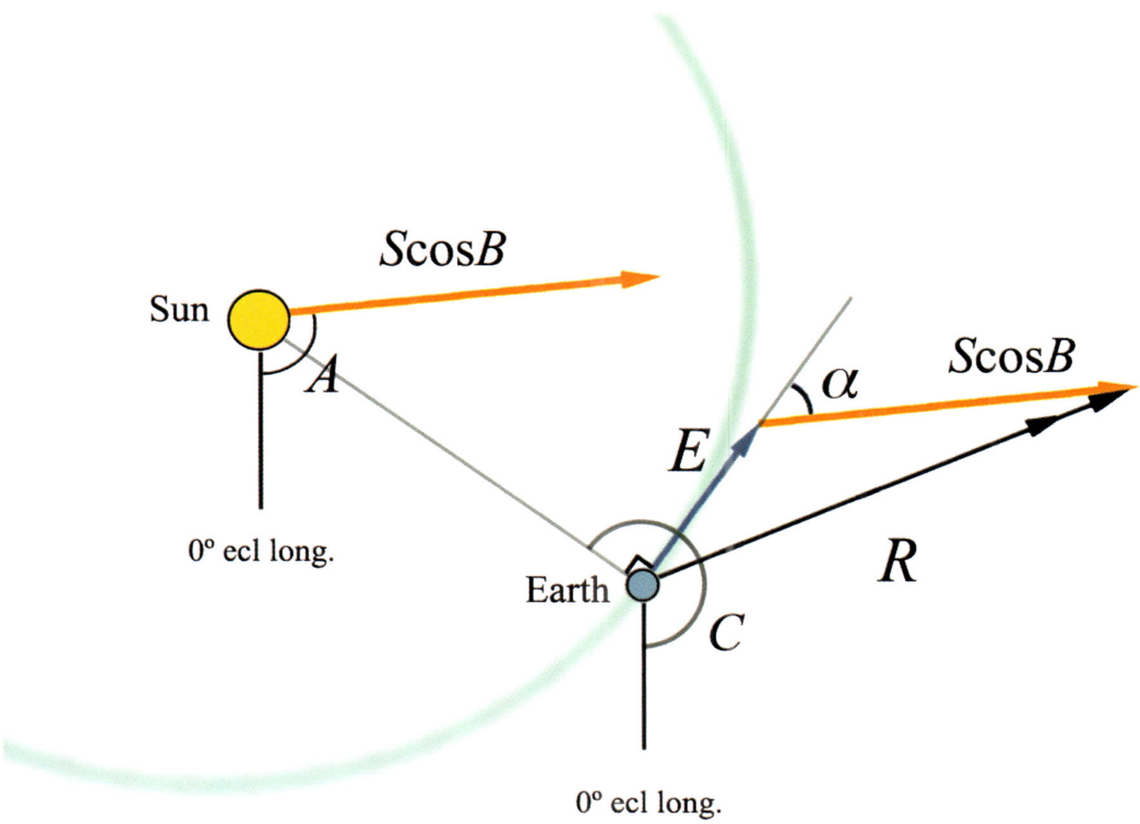

Figure 4.12 Showing the Doppler shift from light emitted by distant sources, observed from the Earth, in the ecliptic plane only.

4.12 Estimation of the effect of a contribution to the Zodiacal Light from sources distant from the Solar System

4.12.1 Computation

Supposing that the continuum spectrum from background starlight contains some kind of absorption feature at 5183.6 Å, it is possible to estimate the 'shift versus elongation' curve which would arise from such radiation as observed from the Earth. The shift of the line, corresponding to a relative radial velocity, depends on the combination of the Sun's proper motion in the local stellar group, and the Earth's orbital motion. Since all the emission takes place at a large distance, no line-of-sight integration is needed, and a further simplification is introduced if we are interested only in the plane of the ecliptic. Only components of velocity in the ecliptic plane are considered here.

We require the apparent radial velocity of distant matter as a function of elongation, which is simply the negative of the ecliptic plane component of the Earth's total motion in the direction specified by this elongation.

The geometry is a simpler version of the continuous flow model, and is shown in Figure 4.12 (previous page). If the Sun's proper motion is S km/s and the direction of the apex of this motion has ecliptic latitude B, the numeric velocity component in the ecliptic plane is $S\cos B$, directed towards ecliptic longitude A. Coordinate conversions from Right Ascension and Declination to ecliptic longitude and latitude were made using tables. For any day in the year the ecliptic longitude of the Sun, C degrees, is given in the Astronomical Ephemeris. The ecliptic latitude is, of course, zero. From this, the direction of the Earth's motion, at speed E km/s relative to the Sun, is $C-90$ (degrees ecliptic longitude). These velocities are shown as vectors in Figure 4.12 in which the angles A and C are measured relative to 0 degrees ecliptic longitude.

The angle, α degrees, between the two components of velocity is

$$\alpha = (C-90) - A$$

i.e. $$\alpha = C - A - 90.$$

Using the cosine rule for the component triangle, since its obtuse angle is 180-α, the magnitude R, of the component in the ecliptic plane of the Earth's total velocity, is given by

$$R = \left\{E^2 + S^2\cos^2 B + 2ES\cos B\cos(C-A-90)\right\}^{1/2}$$

This is directed towards ecliptic longitude L where

$$L = A - \sin^{-1}\left\{\frac{E}{R}\sin(C-A-90)\right\},$$

obtained using the cosine rule in the same triangle.

The elongation of this vector from the Sun

$$\varepsilon_R = L - C$$

and so
$$\varepsilon_R = A - C - \sin^{-1}\frac{E}{R}\sin(C - A - 90).$$

Values of E and S were taken from *Astrophysical Quantities* and R and ε_R were calculated for October and April. The apparent component of velocity of distant matter at an elongation ε from the Sun is evidently

$$-R\cos(\varepsilon - \varepsilon_R).$$

The wavelength shift $\delta\lambda$ for a given radial velocity is found by the usual formula for optical Doppler shift at a wavelength λ

$$\frac{\delta\lambda}{\lambda} = \frac{v}{c}$$

where v is the radial velocity and c is the speed of light. Thus we have

$$\delta\lambda_{distant\ matter} = -\frac{\lambda}{c}R\cos(\varepsilon - \varepsilon_R).$$

This function is shown plotted in Figure 4.13 for the two periods of observation. The figures are

Time	$-\lambda R/c$ (Å)	ε_R
October 30th	0.3	73 W
April 15th	0.6	97 W

The Earth's velocity – approximately 30 km/s – is the dominant factor in the Doppler shift since $S\cos B$ is only about 12 km/s. This means that the maximum blue shift is always on the morning side within about 15 degrees of the 90 elongation position. The Sun's velocity *adds* in the case of the April position, which gives near maximum amplitude of the curve. October is near the minimum case and shows a small amplitude.

4.12.2 Comparison of distant emitting model with the data

Comparison of the curves with our experimental points shows that the rough general shape of the experimental curves can be produced by such a theory, so it is certainly worth finding out if galactic starlight does contribute to the spectrum observed, and in what proportion.

In the October case, the curve is quite similar to that predicted by the rotating dust cloud model but, like the latter, does not satisfactorily fit the data, predicting a cross-over point nearly 60 in error, and a maximum blue shift at an elongation where the experimental points are at around zero shift. The contribution of galactic light in the October observations must thus be small.

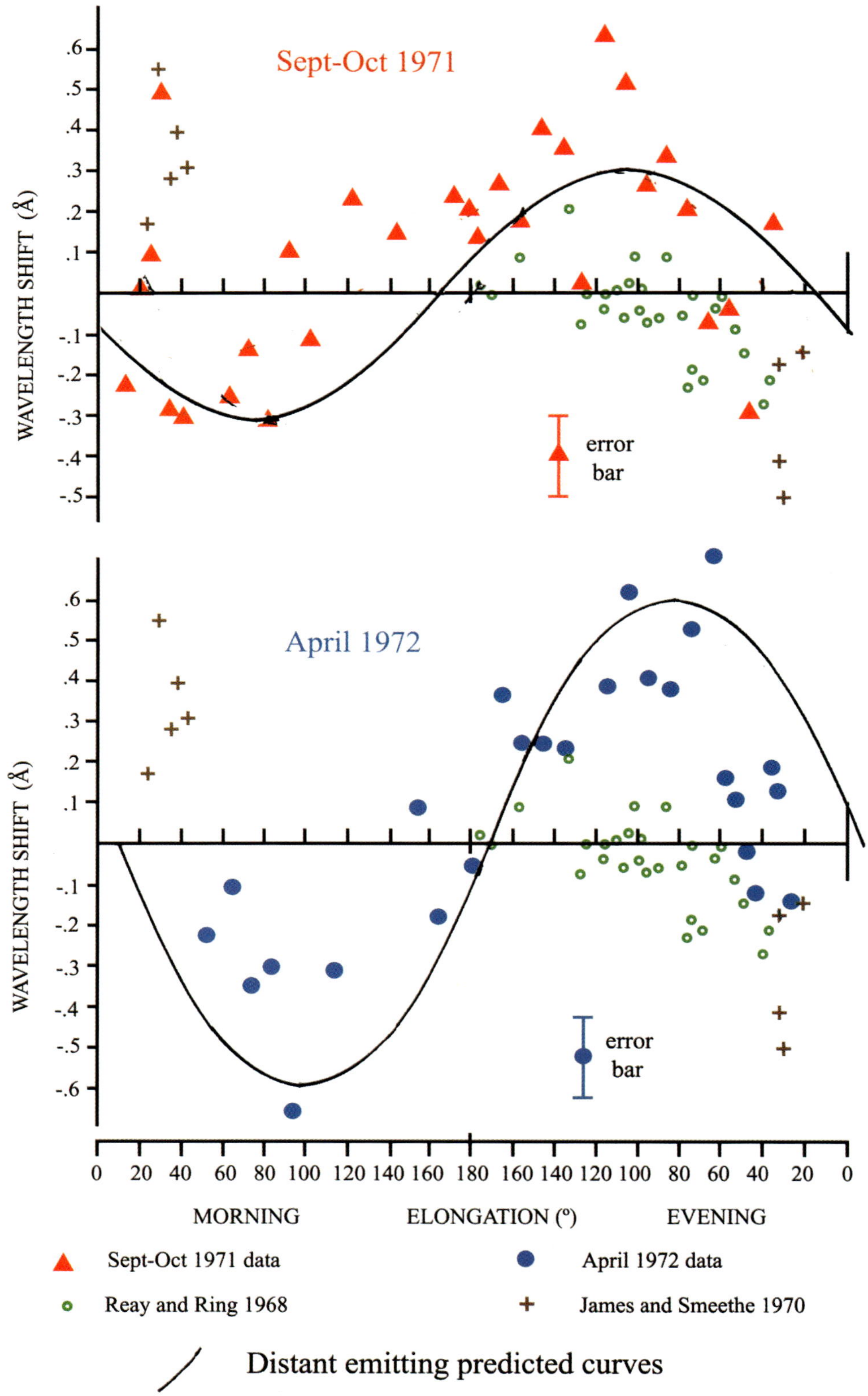

Figure 4.13 Showing all the wavelength shift data available in 1974, compared with predicted curves for distant emitting matter.

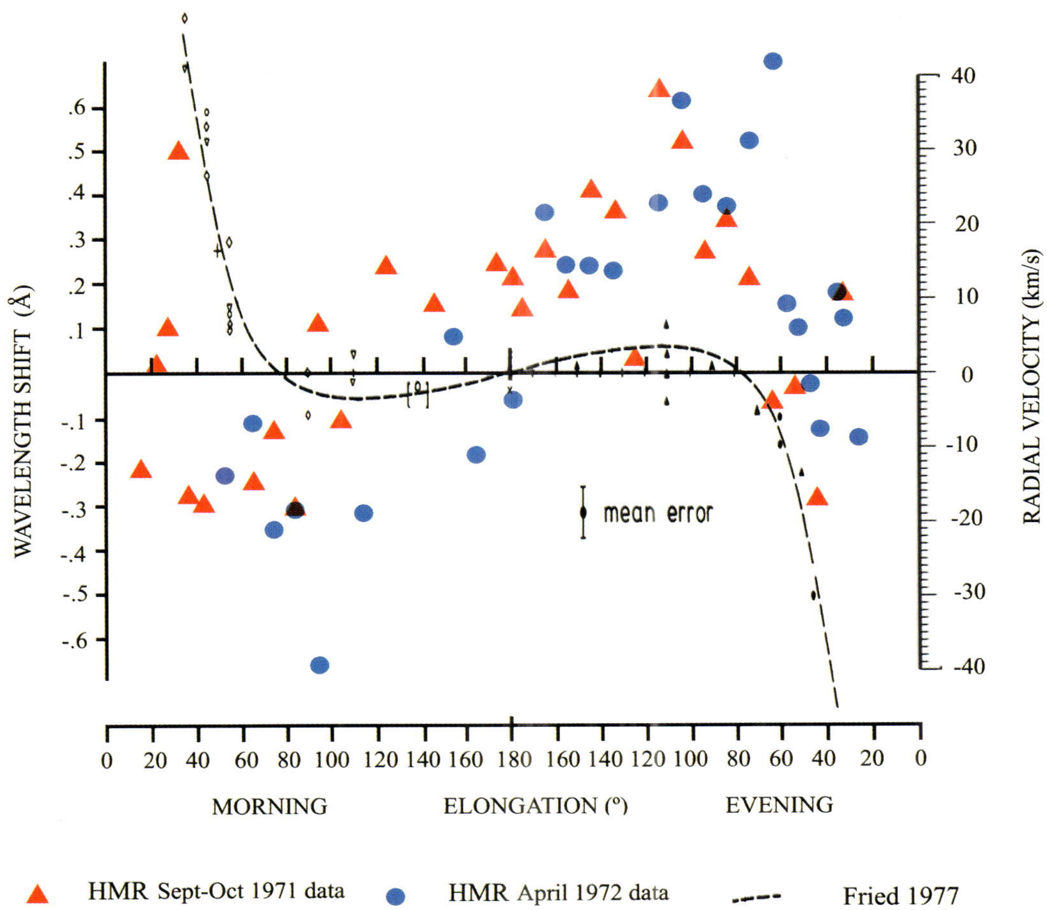

Figure 4.14 Comparison of our data with that of Fried 1977.

In the April case, this theory predicts very large maxima, not commensurate with most of the data. A few points are, in fact, found with shifts up to 0.6 Å, and it is possible that the spectra for these points were measured when the direction was in line with a region of strong starlight background. The outskirts of the Milky Way may have been responsible. Overall, however, the curves fitted to the data do not on the whole agree well with the predictions of this theory. This, added to the fact that considerable care was taken to avoid making measurements in areas of the sky contaminated by other sources, leads me to believe that the contribution of a background starlight absorption line is small enough is most cases to be ignored in our analysis.

This opinion is strongly reinforced by the experimental appearance of line shapes. These are very near to the theoretical predictions of the program CONVOL, which simulates the output when a pure ZL spectrum is input to the system. Diffuse starlight is by no means entirely composed of light from stars with pronounced Fraunhofer features in their spectra. Even if it were, in a particular direction, the lines should be smeared by the combination of proper motions; and so an absorption line, if it exists at all in the spectrum of distant aggregate starlight, would be expected to be shallow and wide. The very fact that our observed shapes are not significantly shallower than the predictions from local dust particle models suggests again that background contributions are, on the whole, small. It is worth noting here that for the future it would be useful to confirm this view of the spectral content of distant starlight, both in theory, in different populations of stars and regions of nebulosity, and in practice, by direct spectroscopy of such regions, without the interference of the ZL. Of course, this requires a viewpoint outside the Solar System.

4.13 Further analysis of the data by comparison with new radial velocity data and model theory available in 2007

As discussed in Chapter 1, there have been long periods of inactivity in Doppler studies of the ZL since 1974. Nevertheless, some interesting results have now been added to the data in this field, and it is very informative to study these for the first time in relation to the complete HMR observations. I present them here in chronological order.

4.13.1 Observations of Fried 1977

Figure 4.14 (previous page) shows the results of Fried (1977). It is immediately apparent that there is little or no agreement here between the two sets of measurements. Reference to Figure 4.5 suggests that Reay's model with $p = 2$ is a fairly good fit to Fried's points, which paints a picture of large numbers of large dust particles rotating in prograde orbits, at speeds close to the orbital speed of the Earth in the case of particles at around 1 AU, giving those very low wavelength shifts around 90 degrees elongation. Indeed Fried himself deduced an ensemble of dust particles in hyperbolic orbits, with a size dependence on orbital radius r given by

$n(r) = r^{-1.3}$, and a corresponding dependence of velocity

$$V(r) = r^{-1.5}$$

If these data had come to my attention in 1977, I would doubtless have felt depressed, since they do not in any way support our results. It is therefore something of a comfort that as time went on, no other observers' findings conformed with these of Fried. This is not, of course, to say that there is anything wrong with these observations. There is still, as we have discussed, debate about the variability of the ZL, and it may well be that, since no-one else was observing its Doppler signature in 1977, Fried saw a temporal effect.

4.13.2 East and Reay (1984)

Figure 4.15 shows the points published by East and Reay (1984) on the single wavelength shift versus elongation graph, plotted in direct comparison with the HMR data.

These data have an improved signal-to-noise over ours – the product of a more stable, servo-controlled Fabry-Perot etalon, and a greater number of observations to average. Overall they show a remarkable congruency with the HMR data, with the qualifications that the wavelength shifts are slightly smaller in the range 60° East to 60° West, and the morning-evening asymmetry is even more pronounced. East and Reay, from a polynomial curve fitted to these results, reported a confirmation of a predominantly prograde orbiting Zodiacal Cloud, with no evidence of retrograde component, in elliptical orbits, rather

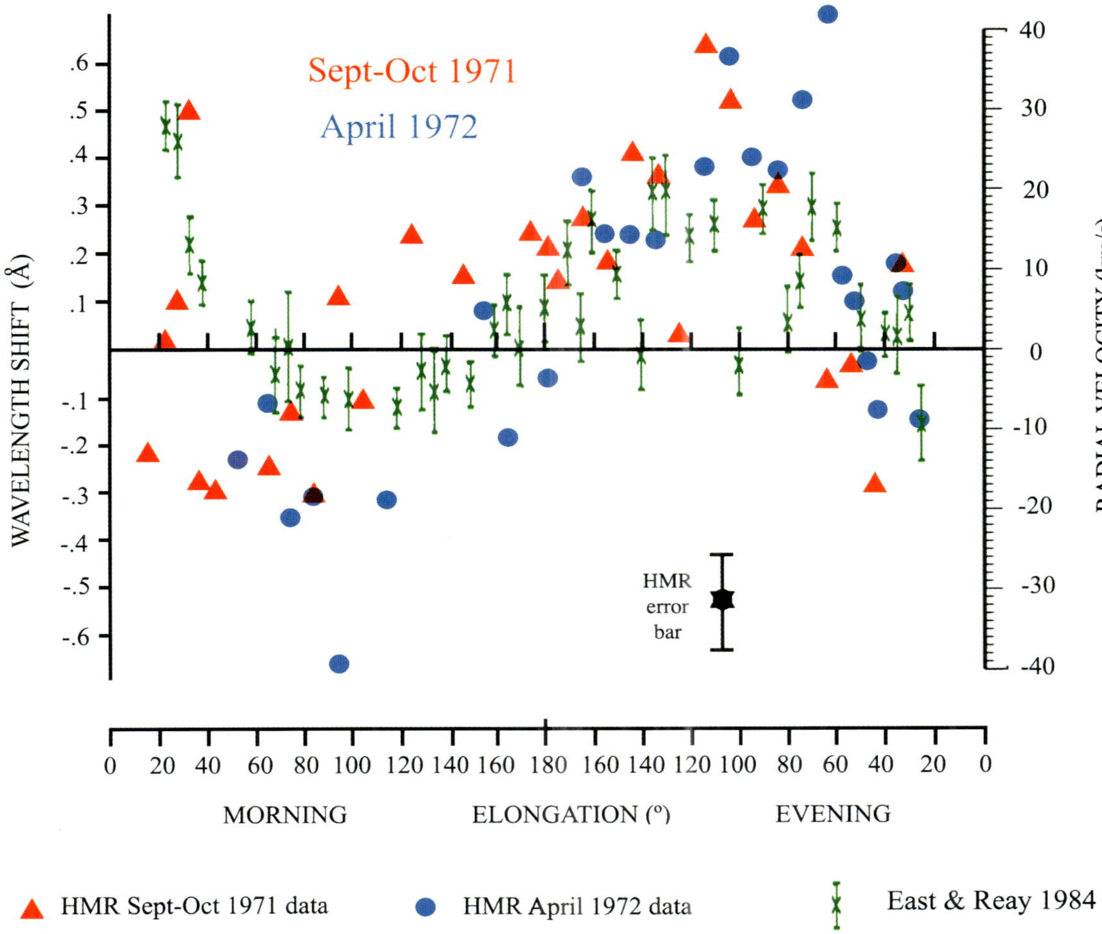

Figure 4.15 Wavelength shift vs. elongation: the results of East and Reay (1984), compared with the Hicks, May and Reay data.

than hyperbolic, and a recession of the Gegenschein of about 2.5 km/s. The Gegenschein results were supported by an extra body of data specifically amassed to reduce errors.

4.13.3 The WHAM observations, and theoretical model predictions

In 2004, Reynolds, Madsen and Moseley published some new data, collected using the WHAM spectrometer, designed for H Alpha mapping, at Kitt Peak, Arizona. The spectrometer itself was a dual Fabry-Perot, 15 cm in diameter, of superior resolution (0.2 Å) and high throughput. The results were published in Reynolds *et al* (2004), and analysed with reference to models by Ipatov *et al* (2006) in Madsen *et al* (2006). A great improvement of signal-to-noise ratio over our measurements was achieved by feeding the spectrometer from a 0.6 m telescope, the whole assembly being remotely controlled from Madison, Wisconsin.

The observations were obtained quickly, on just two nights in November 2002. Because of the higher resolution, the WHAM team were able to identify, and subtract, non-MgI emission lines from the atmosphere, contaminating the MgI absorption line, again improving the reliability of the data. It is

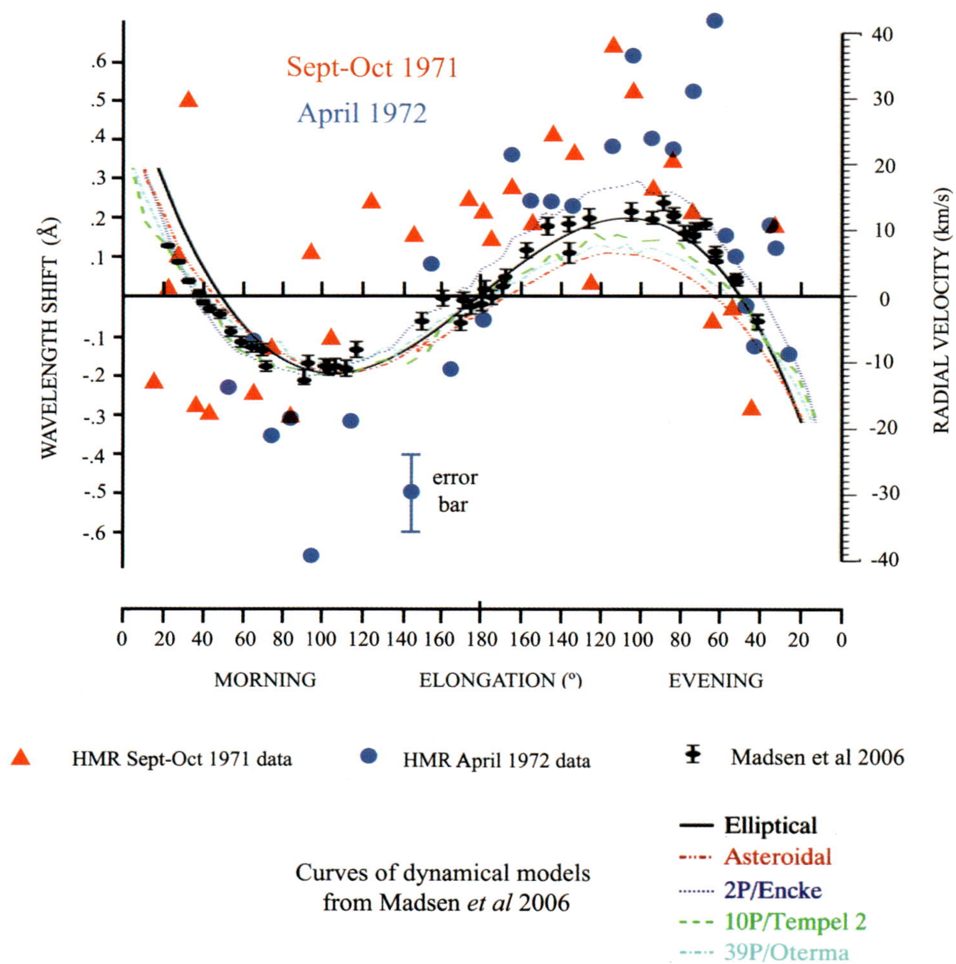

Figure 4.16 Wavelength shift data and theoretical curves from Madsen *et al* (2006), plotted in comparison with HMR data.

possible that these unseen emissions had an influence on our data points – perhaps even producing the morning asymmetry and the red shift of the Gegenschein, though this would be hard to prove. The WHAM results, shown in Figure 4.16, are shown plotted alongside the HMR data. The general agreement between the two sets of observations is quite good, although there is a tendency for the HMR data to present higher shift at intermediate longitudes, and the WHAM data does not show any morning-evening asymmetry, or any shift at 180° for the Gegenschein.

In 2006, Ipatov *et al* published their model of a Zodiacal Cloud, starting from the erosion of comets and asteroids, as mentioned in Chapter 1, and developing orbital velocity elements by simulating their dynamical behaviour with time. Some of the resulting curves are shown in Figure 4.16, and it can be seen that they fit the WHAM observations well. I decided to see how the HMR radial velocity data would look, as compared to the Madsen *et al* (2006) curves, if they were binned into larger groups of averaged scans (next section).

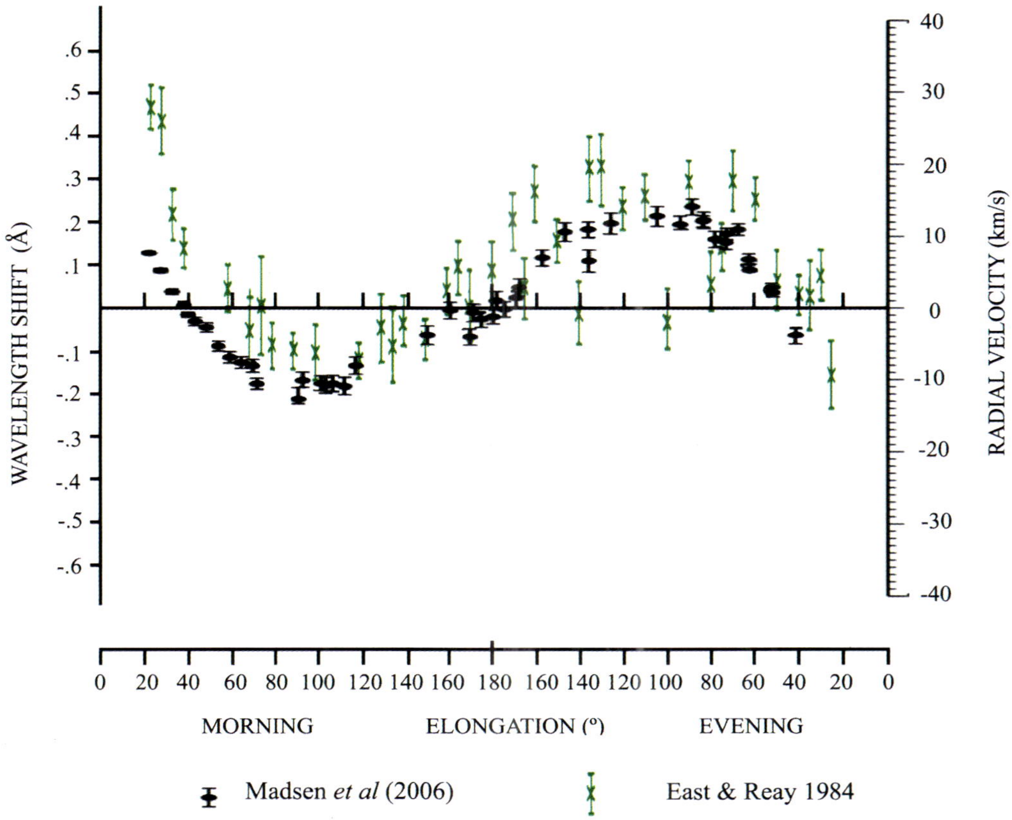

Figure 4.17 Comparison of East and Reay (1984) with WHAM data.

We have already seen the WHAM line-width data in Figure 4.4, and remarked that there was some similarity with HMR in the grouping of the points. Certainly the disagreement in the *wavelength shift* data between HMR and WHAM is not very big, perhaps within the probable errors of the HMR data; but reconciling WHAM with the East and Reay data, with its smaller error bars, is not so easy (Figure 4.17).

The evidence is clearly conflicting here; East and Reay's data shows a distinct divergence from WHAM at all elongations which is not inside the predictions of probable error. Since our HMR data conforms more closely to East and Reay than WHAM, there is still evidently a need for a new set of data, to determine if the differences are a conflict, or the result of seasonal or other variations. The East and Reay results, in almost every case, show a shift towards the red as compared with Madsen *et al* (2006). As we saw earlier, the only model which has a chance of explaining such asymmetries between morning and evening, or Spring to Autumn, is my continuous flow, interstellar dust model. Could it be that we are seeing a combination of rotating and free-flowing dust in the ZL? If the difference between Spring and Autumn radial velocities suggested by our data is real, it is possible that the WHAM experiment, a snapshot in time at the November configuration, was simply fortunate in catching the radial velocities in a neat, symmetrical pattern, perhaps in a period of low density in the interstellar dust component. Perhaps only time and still higher resolution and signal-to-noise will tell, but there would clearly be

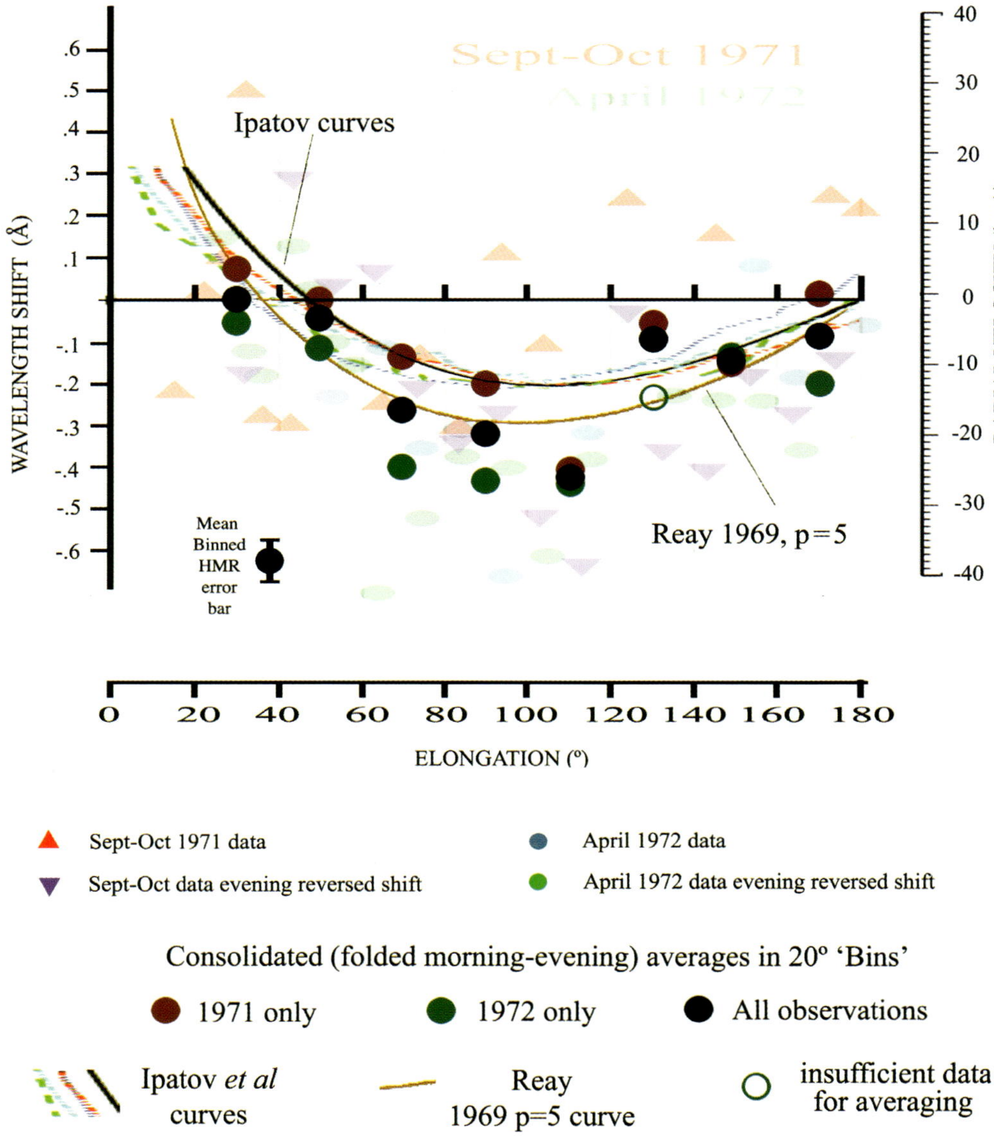

Figure 4.18 The result of binning the HMR data in wavelength intervals of 20°.

good reason to try to monitor the radial velocity versus elongation curve all the year round. There is one more piece of reductive analysis on the HMR data which can be done: binning.

4.14 'Binning' of the HMR data

Bins of 20-degree spans in wavelength were chosen, and the mean shifts in each 20-degree interval were calculated. The results are shown in Figure 4.18.

This figure shows only elongations from zero to 180°, because the plotted points for evening and morning have been 'folded' together, reversing the sign of the evening observations, as many previous authors have done, assuming the shifts to be symmetrical about the antisolar point. The 'ghost' points

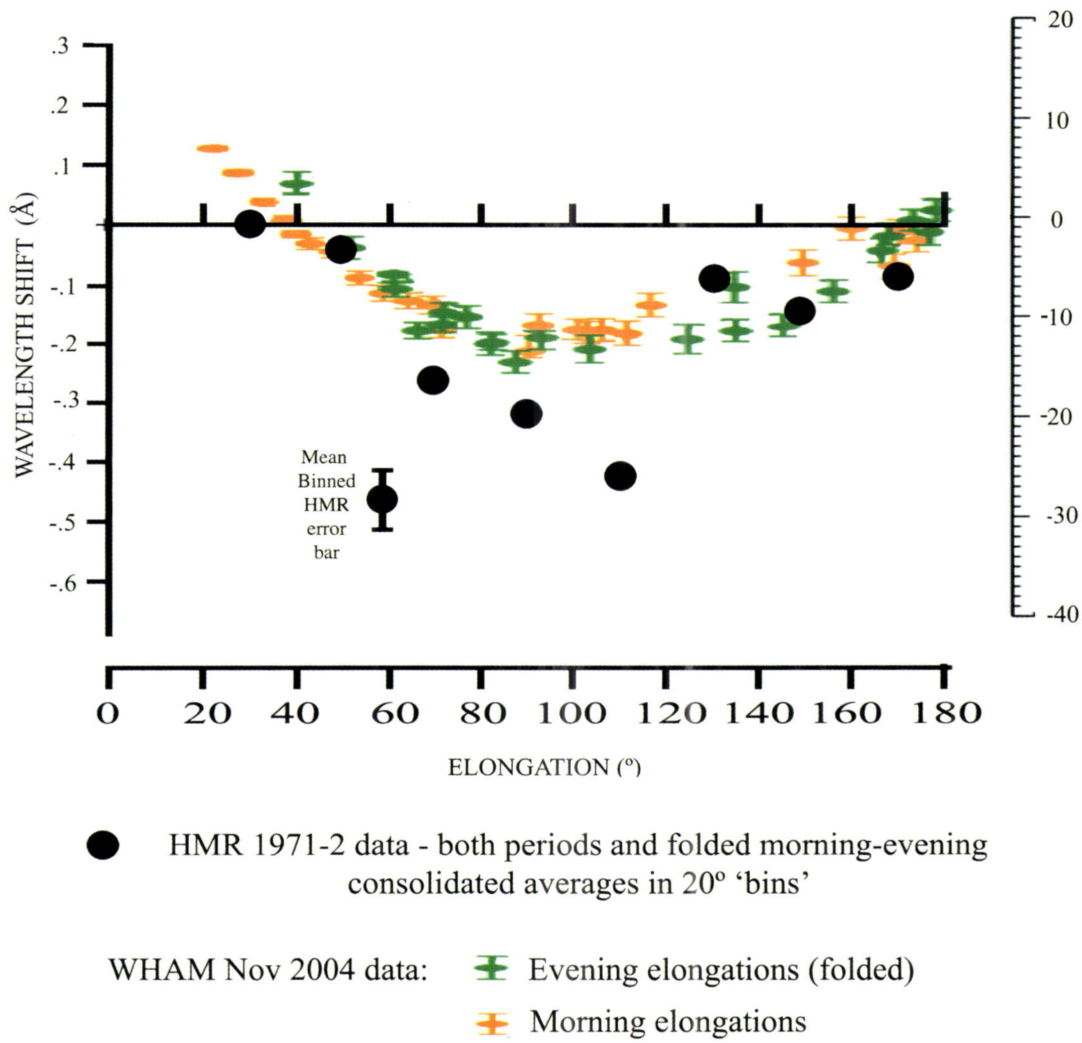

Figure 4.19 Binned folded HMR data, averaged in wavelength intervals of 20°, compared with folded WHAM data.

are the original HMR data. Three sets of binning averages are shown, the results for 1971 only, for 1972 only, and for both sets averaged together. Except for the region 110°-130°, the HMR observations consistently show a shift about 5 km/s larger than the Madsen/Ipatov curves. Again the fit seems quite good for a Reay curve in which $p = 5$. The separate binned HMR points underline the view that there is a real difference between morning and evening radial velocities, and a time dependence.

Finally, Figure 4.19 shows a direct comparison between the consolidated HMR data and the WHAM data, both folded about the antisolar point. The folding reveals that there actually *is* a morning-evening asymmetry in the WHAM points – not immediately apparent in the un-folded figures. The morning points seem to cross the zero line at about 30° elongation, and the evening points at about 50° – a discrepancy well outside the limits of the predicted uncertainties in the data. The HMR data shows a crossing at about 40°, but is lacking in enough low elongation data to be sure. Elsewhere the agreement between the two sets of data is good except between 60° and 110° - where our radial velocities are almost twice that indicated by the WHAM data. It has to be emphasised that all the information on time variation, if there is any, has been eliminated from this graph.

4.15 Variations with time

I did not expect to find evidence of short-term changes in the parameters of the Zodiacal Cloud. But the differences between our two periods of observation (at opposite ends of the Earth's orbit) are so pronounced that it is plainly worth checking in the future to see if the differences can be reproduced. Throughout the history of the observation of the ZL there have been reports of seasonal and other variations in intensity of both morning and evening ZL. But there has not been a consensus on what these variations are, or the magnitude of the changes, and since very few truly long-term projects have been undertaken, it has not been established whether the fluctuations are cyclic or not with the season, or due to unique events, such as solar flares or visits from comets, or indeed, as discussed later in this Chapter, due to influences beyond the Solar System.

Variations in intensity were recorded by naked-eye observations as long ago as 1856, by G. Jones in his voyage from the USA to Japan and back on a boat, which I mentioned in Chapter 1 (Figure 1.1). Jones's meticulously drawn nightly contour maps of the Zodiacal Band show what he thought was a dependence of the ecliptic latitude of the cones on the latitude of the observer. Jones interpreted this dependence as an indication that most of the Zodiacal Cloud was attendant on the Earth, rather than the Sun. Though this became an unfashionable view in the second half of the 20th Century, the existence of a component of 'local' light-scattering dust can still not be ruled out, especially since the discovery (Reach *et al* 1995) of the 'Earth Tail' of dust, gravitationally channelled by the passage of the Earth on its journey around the Sun. It is also possible, however, that the variation Jones saw was *not* latitudinal, but seasonal, and this might turn out to be the earliest observed manifestation of the departure of the ZC from the invariable plane, which has been documented in recent satellite experiments (Hauser *et al* 1984).

E. O. Hulbert (1930), in a much later analysis of the Jones observations, wrote that Jones had witnessed 23 unusually bright periods of the ZL, 16 of which took place 3 days after a magnetic storm. Hulbert also noted other enhancements preceded by magnetic activity. He notes, however, that some of the variations might be interpreted as overall sky brightness changes, as a true separation of ZL component does not seem to have been made. It seems likely to me that Jones himself would have placed more significance in his observations of the variations *shape* of the ZL than those of overall intensity.

In more recent times, seasonal oscillations in brightness, of magnitude about 20 per cent, were reported by Elvey and Roach (1937) using photoelectric measurements.

In the evening ZL they found a maximum in April-May, and a minimum in January-February; and an opposite effect in the morning ZL. These conclusions were supported by their reduction of visual Japanese observers.

Divari (1955), found seasonal variations in which the morning ZL intensity increased in September to November, and the evening intensity decreased from January to March. These results clearly do not concur with Elvey and Roach's. As noted in Chapter 1, Sanchez (1974) found a correlation between ZL brightness and solar activity.

The question of variability of the ZL seems central to interpretation of the Doppler data so far gathered in 2007, to account for the discrepancies revealed in this thesis between all currently available sets of results.

4.16 Summary of the implications of the 1970–4 work, and pointers for the future

It is clear that a reasonable fit to our data can be obtained using a model of the Zodiacal Cloud comprising dust particles, some in orbit around the Sun, and some drifting through the Solar System, possibly to be identified with interstellar dust. The differences between the two periods of observations, if they are real, may be attributed either to a seasonal effect, corresponding to an asymmetry of the dust cloud with respect to the Earth's orbit, or to short-term fluctuations of the orbiting or drifting component. Such short term variations in the former might be due to the visits of comets, or variations in Solar Wind. In the latter, the differential would indicate either that there are short-range inhomogeneities in the interstellar medium, or that, in its passage through the Solar System, the flow of such material is disrupted, for instance by the planets themselves.

In future work the data from high elongations may be the most useful, since a pure rotating model can account for the shape in all cases for which $\varepsilon \geq 70$ (degrees of elongation from the Sun). Our results are consistent with the theory that a higher percentage of drifting dust was present in September-October 1971 than in the April period the following year.

More data is still required, critical for distinguishing between models, and hence for belief in the conclusions we are drawing, and more precise determinations of the actual spectrum profiles, rather than simply averaged shifts, will in the future yield better mapping of the motions of the interplanetary dust cloud.

Chapter 5
Current developments and future plans

In this chapter, I will briefly review the experiments currently planned, or in operation, in Solar System Astronomy: rocket-borne, space-borne, and ground-based observations, which are likely to throw new light on the structure and nature of the Sun's dust cloud. I will also touch on present unresolved issues which warrant attention, and sketch plans for future observations which Dr. Garik Israelian and I plan to make, to extend and refine the Doppler radial velocity measurement technique. We hope to map dust kinematics to add new detail to the latest picture of the Zodiacal Cloud, as revealed by IRAS, COBE and ISO.

5.1 Current Zodiacal Light research

CIBER

The Cosmic Infrared Background Experiment (CIBER) (Bock *et al*, 2006), is a rocket-borne instrument designed to search for signatures of primordial galaxy and star formation in the cosmic near-infrared extra-galactic background, in the wavelength region 0.8 to 2 microns. Its launch, funded by NASA, is scheduled for September 2007. It is hoped that CIBER will settle the question of whether the DIRB really *is* the echo of the first stars and galaxies. CIBER consists of a wide-field two-colour camera, a low-resolution absolute spectrometer, and a high resolution narrow band imaging spectrometer. It will trace the Zodiacal Light with the low resolution spectrometer, using the *intensity* of scattered Fraunhofer lines, to provide an independent measurement of Zodiacal emission, and a new check on the Kelsall *et al* Zodiacal dust models, which were used to compare with the results of the Diffuse Infrared Background Experiment (DIRBE) of the Cosmic Background Explorer (COBE). It is always necessary to subtract an accurate estimate of the intensity of ZL emission in order to map the Galactic contribution and the level of emission from elsewhere – the Infrared Background. Models of the levels of ZL contribution are still so uncertain that estimates of the IRB are, at present, tentative. Being essentially a low resolution experiment, CIBER will not be able to do spectroscopy, as such, of the ZL. But the CIBER team hope to accurately assess the contribution of local (z = 1-3) galaxies, and hence to determine the excess unaccounted for. In this residual radiation, CIBER will be searching for the newly-predicted signature of high luminosity primordial stars and mini-quasars - a red-shifted Lyman cut-off feature between 0.8 and 2 microns.

PLANCK

It will be interesting to see what view the Planck experiment, due for launch in 2008, will have of the Zodiacal Light. The satellite will be put into a Lissajous orbit around the Sun-Earth libration point L_2, approximately 1 AU from the Earth, which will give it a significantly different perspective on the Zodiacal Cloud. And, 'seeing' at wavelengths over 300 microns, it is possible that the ZL will look very different in shape. Planck will construct all-sky maps in nine frequency bands, covering the range 30 GHz to 857 GHz, with relatively high sensitivity and resolution (Maris *et al,* 2006). According to the Planck 'Bluebook' (ESA, 2005), Planck is expected to detect Zodiacal Light at 857 GHz and possibly 545 GHzd; at these wavelengths the Zodiacal emission is, until now, poorly constrained by observations, and its level uncertain by more than a factor of two. At lower frequencies than this, levels of emission from the ZL are expected to be very low. It is hoped that accurate mapping of the dust emission in this wavelength region will give information on the geometrical distribution of the largest (up to 1 cm) grains in the dust cloud, in three dimensions, and as a function of season. The detection of thermal emission at wavelengths as long as this makes Planck sensitive to dust colder than has been previously mapped; finding such emission could imply the existence of an extension of the Zodiacal Dust Cloud into the outer reaches of the Solar System, perhaps into the Kuiper Belt. Of course, even if the cloud only extends to the asteroid belt, there would be Rayleigh-Jeans radiation from warmer dust at 350 microns. The signature of colder, more distant dust would be an enhanced intensity of submillimetre emission. The possibility of Planck detecting the IRAS dust bands does not appear to have been discussed; however it is hoped that it will be able to map the cometary dust trails reported by IRAS.

SPITZER

The Spitzer orbiting infrared telescope was named after the first astronomer (Lyman Spitzer in 1946) to suggest putting a telescope in space. It has a sensitivity 1,000 times better than ISO, and has full wavelength coverage in the mid-IR (Werner *et al*, 2004). The *"First Look Survey"* of Spitzer (FLS) has investigated the mid-IR Background brightness in the ecliptic plane. The sensitivity of Spitzer is background-limited, and dominated by Zodiacal Light at all but the longest wavelengths and lowest Galactic latitudes. Spitzer carries a short wavelength Infrared Array Camera (IRAC), which provides absolute photometry, relative to the brightness of a dark sky patch, at a range of four wavelengths spanning 3.6 to 8 microns. This instrument is sensitive to emission from evolved galactic systems, and has produced many beautiful multi-'colour' images of many nearby galaxies. The 24 micron long-wavelength Multiband Imaging Photometer for Spitzer (MIPS) has provided absolute photometry of ZL emission. The measured ZL background was found to be within 5 per cent of Spitzer model predictions at 24 microns derived from COBE/DIRBE data. The total background brightness at 8 microns was 36 per cent brighter than the model predictions. However, this is thought to be consistent with known intrinsic variations of the ZL as a function of viewing geometry (Meadows *et al* 1997).

Recent reports of analysis of Spitzer data (Reach 2005) indicate that it is indeed achieving success in the pursuit of refinement in the known structure of the Zodiacal Cloud. They say, *"Using the new infrared spectral and imaging capabilities of the Spitzer space telescope, we find dust trails from short-period comets, new structures in the zodiacal light, weak 9-11 micron silicate spectral features from some comets and the zodiacal light, and the color temperature of cometary nuclei, comae, and dust trails. The high frequency of dust trail detections for short-period comets suggests they are a generic phenomenon and that most such comets lose mass primarily in the form of meteor-sized particles. The color temperature of the trails is higher than a grey-body, while their dynamics suggest they must be of order mm-sized, so they must maintain temperature gradients across their small surfaces. The zodiacal light silicate feature requires a contribution from small amorphous silicates, though the continuum requires larger particles."* This quote has direct relevance to the work I hope to undertake in the coming year.

5.2 Objectives of our proposed new work in Zodiacal Light

5.2.1 Doppler measurements on scattered light at optical wavelengths

Dr. Garik Israelian (of the Institute of Astrophysics, La Laguna, Tenerife) and I believe it would be very instructive to look at Doppler shifts in regions away from the ecliptic plane, and see if we can detect a higher proportion of retrograde material. We have seen that high radial velocities might also be an indication of a significant Interstellar component. If we can obtain good enough quality spectra in a slice perpendicular to the ecliptic, including the ecliptic pole and the Gegenschein, it should be possible to choose between these hypotheses.

We have seen that the IRAS observations, interpreted by Low *et al* (1984), indicated the existence of dust bands, producing 'shoulders' superposed on the smooth photometric shape of the traditionally known Zodiacal Light. We propose to look for the dust bands' dynamic signature, by measuring Doppler shifts in a slice of the ZL across the ecliptic plane, at an intermediate elongation, again using a ground-based optical and mid-infrared spectrometer. A sudden change in radial velocity at some point in this slice would be confirmation of the existence of such a discrete component in the cloud, and would offer the first opportunity to determine its motion; this could then be compared with the motion of the putative parent asteroidal families.

As for instrumentation, with the benefit of 30 years' hindsight, it seems to me that, as far as maximising flux is concerned, we, that is, all the F-P radial velocity observers of the time, missed the obvious. The Fabry-Perot instrument was used because of its high throughput at a relatively high resolution, but, by using the pre-monochromator, we actually threw away 3,000 Å of information in the visible alone, all of which contained absorption lines shifted by the same amount, and all of which therefore potentially could have given us more information on radial velocity. In admitting only actually 6 Å spectral range, we rejected 99.8 per cent of the available useful signal. In addition, we used only a small coelostat to collect light. The justification was that we wanted to look at a large area on the sky to achieve large flux-levels, and it would have been hard to find a telescope with a comparable field of view, of one degree, on the sky. But in fact larger flux levels can be obtained with large mirrors with a lesser field of view, more than compensating for this gain. In future I would like to correlate the modification of the *whole* spectral range available, and thus vastly improve the signal-to–noise in the observed spectra. Moreover, with modern echelle technology, there is every reason to employ the largest collector of light available to feed the spectrometer in future. These two improvements ought to allow not just comparison of an 'average shift' with theory, but detailed comparison of multiple spectral line shapes with theoretical models, yielding a much more thorough and accurate mapping of radial velocities. Exploratory experiments will be made in July 2007 with an echelle spectrometer attached to a 3.5 m telescope to test the validity of these arguments and see what quality of results can be obtained.

Recently, Bernstein, Freedman and Madore (2002) detected extragalactic background light using an echelle spectrograph at the 2.5 m Du Pont telescope in Las Campanas (Chile). These authors have developed a technique that uses the strength of the Zodiacal Fraunhofer lines to identify the absolute flux of the Zodiacal Light in the complex night sky spectrum. The analysis is much simpler if we are only interested in the Doppler shifts of Zodiacal Fraunhofer lines.

Dr. Israelian and myself are planning observations with the SARG spectrograph located at the Nasmyth focus of the 3.5 m *Galileo* telescope in La Palma, Canary Islands. SARG is a high efficiency echelle spectrograph designed for the spectral range 370 to 900 nm, and for resolution from R=29,000 up to R=164,000. Both single object and long slit (up to 30 arcsec) observing modes are possible. A dioptric camera images the cross-dispersed spectra onto a mosaic of two 2048 by 4096 CCDs. We plan to use a long slit mode (aperture 5, providing an area 1"07 by 26"7 arcsec) which provides a resolving power of 43,000.

To minimize read-out noise, we will bin the data on the chip in intervals of 4 pixels in the spatial direction, and average over the full extent of the slit in the data reduction. We will use 30 minute exposures which are expected (given the results of Bernstein *et al* 2002 and the Exposure Time Calculator of SARG) to provide a signal-to-noise ratio between 60 and 80 per resolution element. We shall use a cross disperser CD1 (600 g/mm, 4840 angstrom, order separation 10.7 arcsec) which provides spectra in the range 3,600-5,140 Å in 50 orders. This spectral window is not severely contaminated by sky glow (Bernstein *et al* 2002), being in the blue part of the spectrum, and we hope to be able to observe ZL lines directly in these spectra without insurmountable problems in sky subtraction.

The solar Fraunhofer lines are clearly visible in this part of the spectrum and they will be Doppler-shifted due to the scattering by dust particles. We will use the Solar High Resolution Spectral Atlas of Kurucz *et al* to identify and study the Fraunhofer lines. We are not interested in absolute flux determinations. This experiment will allow us to increase the precision of Doppler measurements due to the fact that we will be observing many Fraunhofer lines simultaneously. We aim to observe the ZL at various latitudes and elongations, and will study the velocity fields of the dust bands in visible light. Our study will allow us to constrain models of the ZC, and dust band formation.

5.2.2 Doppler shifts of emission lines in the infrared

Investigating the ISOCAM spectra, Reach *et al* (1997) found a hint of a 9-11 micron feature, and proposed that the particles producing the ZL are composed of silicates similar to those found in the tail of P/Halley. The mid-infrared spectra of many comets show structured 9-11 micron emission bands, and the spectrum of the dust cloud around Beta Pictoris has been found to exhibit a comparable structure. The shape of the ZL silicate emission line appears to be similar to those of comet Halley and the circumstellar dust around Beta Pictoris, but the ZL silicate feature is superposed on a relatively brighter continuum. It is very important to confirm the detection of the silicate line at 9-11 microns. Further work with Spitzer may reveal more structure in this feature. We propose to look at it from the ground.

We plan observations of the silicate feature with VISIR on the 8.2 m VLT (UT3) in Chile. Using a high resolution mode (providing resolution 25,000 at 10 microns), we will try to observe in the 8-12 micron atmospheric window. We propose to use a long slit mode: a slit width of 1 arcsec, and slit length 32.3 arcsec. The sky (airglow) lines will be subtracted by the "dithering" and "nodding" method (see for example, the Gemini Near Infrared Spectrograph website 2007).

These observations will allow us:
1) to verify the presence of the silicate emission feature.
2) to study the line profile and its variability.
3) to study the silicate feature in ZL dust bands.
4) for the first time, to attempt Doppler shift measurements on an *emission* line, if there is enough fine structure in the feature to allow this.

We feel that, in the future, the kinematic Doppler-shift signatures of moving dust particles will provide a complimentary source of information, alongside continuing photometry, on the structure of the Zodiacal Cloud: this will lead to more confidence not only in our understanding of interplanetary dust and its origins, but also in our assessment of the emissions from beyond – from Interstellar, Intergalactic, Cosmic, and Background sources.

Almost 35 years after my first experiences with a Fabry-Perot spectrometer aimed at the Zodiacal Light, there are still excellent reasons to continue, with ever improving techniques, to map Radial Velocities in the Zodiacal Dust Cloud.

5.3 Concluding remarks

5.3.1 Summary

The results of my work in 1970–4 were, as planned, the first comprehensive survey of Doppler shifts in the Zodiacal Light, covering elongations in the Ecliptic plane from 25° East to 25° West. The spectra and plots of wavelength shift versus elongation stand as a body of work which continues to be cited by those who have followed in the field. There is fair agreement between my results and those of East and Reay, and Reynolds *et al*, who both worked later with higher power equipment. The differences between the bodies of data still pose questions which further observations will be able to clear up: the question of whether there are variations over time in the kinematic structure of the ZL and if so, whether they are seasonal, and apparent, caused by the Earth's journey around the Sun, perhaps through a continuous flow interstellar component of dust, or if they reflect actual structural changes in the Zodiacal Dust Cloud, perhaps due to asteroidal events or visits from comets.

My conclusions at the time were that an emission component at MgI existed, generated in the Earth's atmosphere, that the Zodiacal Cloud consisted of dust particles of the order of 1 micron, in orbit around the Sun, mainly in a prograde sense, and that the morning-evening asymmetry we found might be due to components of dust NOT in orbit around the Sun. We found no evidence for contra-rotating dust, but at the resolution used I do not consider that the presence of a small percentage of such dust is ruled out. The recession of the dust in the area of the Gegenschein was, and is, an unsolved mystery. This effect was confirmed by East and Reay, but not seen at all by Reynolds *et al.* So it is not entirely clear whether it really exists or not, or if it *sometimes* exists! The results of Fried are not congruent with any of the other three sets of data, but apart from these, the rest of my, and our, conclusions of 1974 remain supported by all other subsequent experiments.

5.3.2 Visions

Often it is hard, among the wealth of technical details published on such a subject, to form, in the head, a clear picture of the physical shape of the Zodiacal Cloud – the kind of graphic visual we would like to

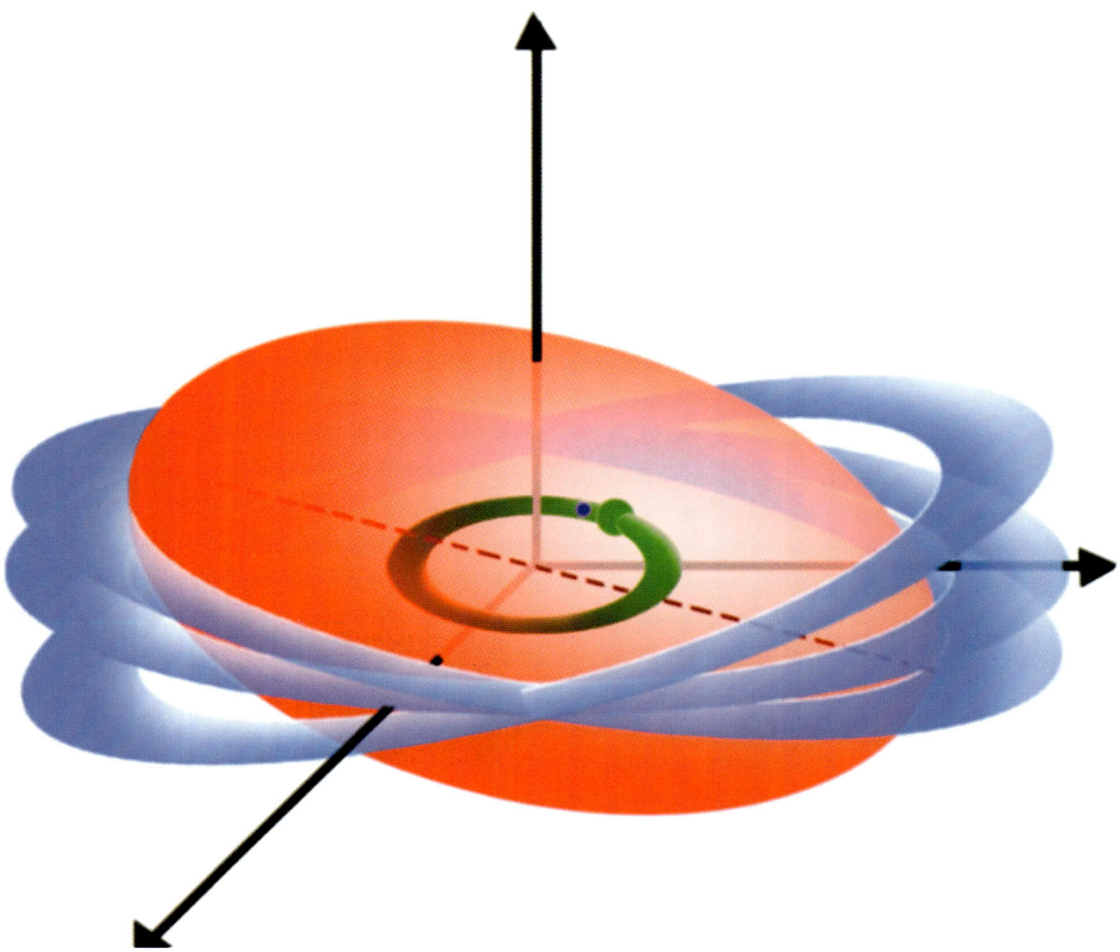

Figure 5.1 From Stark (2006). A geometrical representation of the 1998 DIRBE model of the Zodiacal Cloud. The red spheroid represents the smooth diffuse component (no structure attempted); the blue rings are the three (presumed asteroidal) dust bands; the green ring is the circumsolar Earth-resonant ring, showing a 'Blob' following the Earth (coloured blue).

present to our children, and those who follow us. Figure 5.1, from Stark (2006), is a rare attempt in the published literature to provide a geometrical visualisation of the structure of the dust cloud. The figure is diagrammatical, but clearly illustrates the three dust bands, showing them as flat rings, and makes no attempt to show the 'hollow torus' indicated by evolutionary models.

It is highly likely that this representation will have to be updated, as new and better measurements are achieved.

The final three figures which follow are *'artist's impressions'*. Figure 5.2 is a representation of how the consensus of Zodiacalists pictured the Sun's dust cloud in 1970 – quite smooth and diffuse, roughly lenticular in shape, symmetrical about the ecliptic plane, its density decreasing quite quickly beyond the orbit of Mars. Figures 5.2 and 5.3 are speculative visualisations of the information currently available in 2007.

In Figure 5.3, the various currently accredited components of the dust cloud are illustrated: again a relatively smooth continuous ellipsoidal (lenticular) main cloud, perhaps spreading a little less in the direction perpendicular to the ecliptic than the Fan Model suggests, but now showing significant structure. Each inner planet has a focussing resonant effect, producing a system of circum-solar rings at the radii of the planets - and corresponding depleted areas between the planetary orbits. The innermost rocky planets also produce 'blobs', similar to, but smaller than the blob which has been observed following the Earth, and they have an overall effect, over time, of producing 'lumpiness' in the structure (not yet observed). But most spectacularly, the smooth dust cloud is now flanked by discrete inclined dust bands, originating from collisions in the asteroid belt. This visualisation (Figure 5.3) follows the interpretation of Stark (2006) in Figure 5.1, the IRAS observed photometric irregularities each being attributed to a separate narrow dust band. Some asteroids can be seen, including some of the families which have produced the dust bands, outside the orbit of Mars. There is an additional faint huge, cool dust ring, out towards the Kuiper belt, again, *not yet seen*! The angles of inclination of the rings are depicted schematically only.

Figure 5.3 shows an alternative and probably more plausible interpretation of the IRAS dust band discoveries. Models developed by Dermott *et al* (private communication) show how a single collision between large asteroids can initially produce a band of dust debris distributed along the path of the 'parent' bodies, similar to those shown in Figure 5.3. But over a few million years the models show the individual orbits of the particles precessing at various rates, until eventually the major axes of their elliptical orbits are spread randomly around the Sun, producing a smooth, fat *toroidal* distribution. The orbits of these particles may be thought of as a component in the ecliptic plane, plus a synchronous sinusoidal "North-south" movement. Hence, viewed from the Earth, which is situated inside the torus, the dust particles spend more time at the edges of the 'doughnut' than inside, so we see peaks in optical depth at the top and bottom edges of the torus. Evidently these peaks would be symmetrically placed above and below the ecliptic, which, in the case of the 10° IRAS bands, is certainly what is observed. Figure 5.4 shows one such torus, a highly evolved remnant of an asteroidal collision. The complete picture may require the addition of at least two more toroidal components.

Perhaps some combination of Figures 5.3 and 5.4 would truly represent the appearance of the complex circum-solar dust cloud that will be eventually photographed from a remote space craft in the future. In addition to the asteroidal debris, a comet is seen approaching the Sun, a reminder that quite possibly the major fraction of the dust is actually due to comets. Indeed, to this picture should probably be added a whole fine tracery of cometary dust trails, the ejecta which follow the orbital paths of the comets on their journeys to and from the Sun. And, just possibly, we should add a fairly uniform 'background' of dust flowing through the Solar System, a consequence of the Sun's journey through interstellar space; one wonders how much of a contribution this might be making to current estimates of the Diffuse Infrared Background. If we could accurately map Doppler shifts in this residual drifting component, we would know.

Perhaps in the near future, kinematic studies of the material around us will resume their rightful place alongside photometry, in unravelling the secrets of the formation and evolution of the Zodiacal Dust Cloud.

The 'artist', by the way, is me.

Brian May, 2007

Figure 5.2 (Page 157) Artist's impression of how the Zodiacal Cloud was perceived in 1970.

Figure 5.3 (Page 158) Artist's impression of the currently perceived Zodiacal Cloud, seen, enhanced, from the orbit of Uranus. The Sun is omitted.

Figure 5.4 (Page 159) Showing an impression of an evolved toroidal dust band resulting from a single asteroid collision.

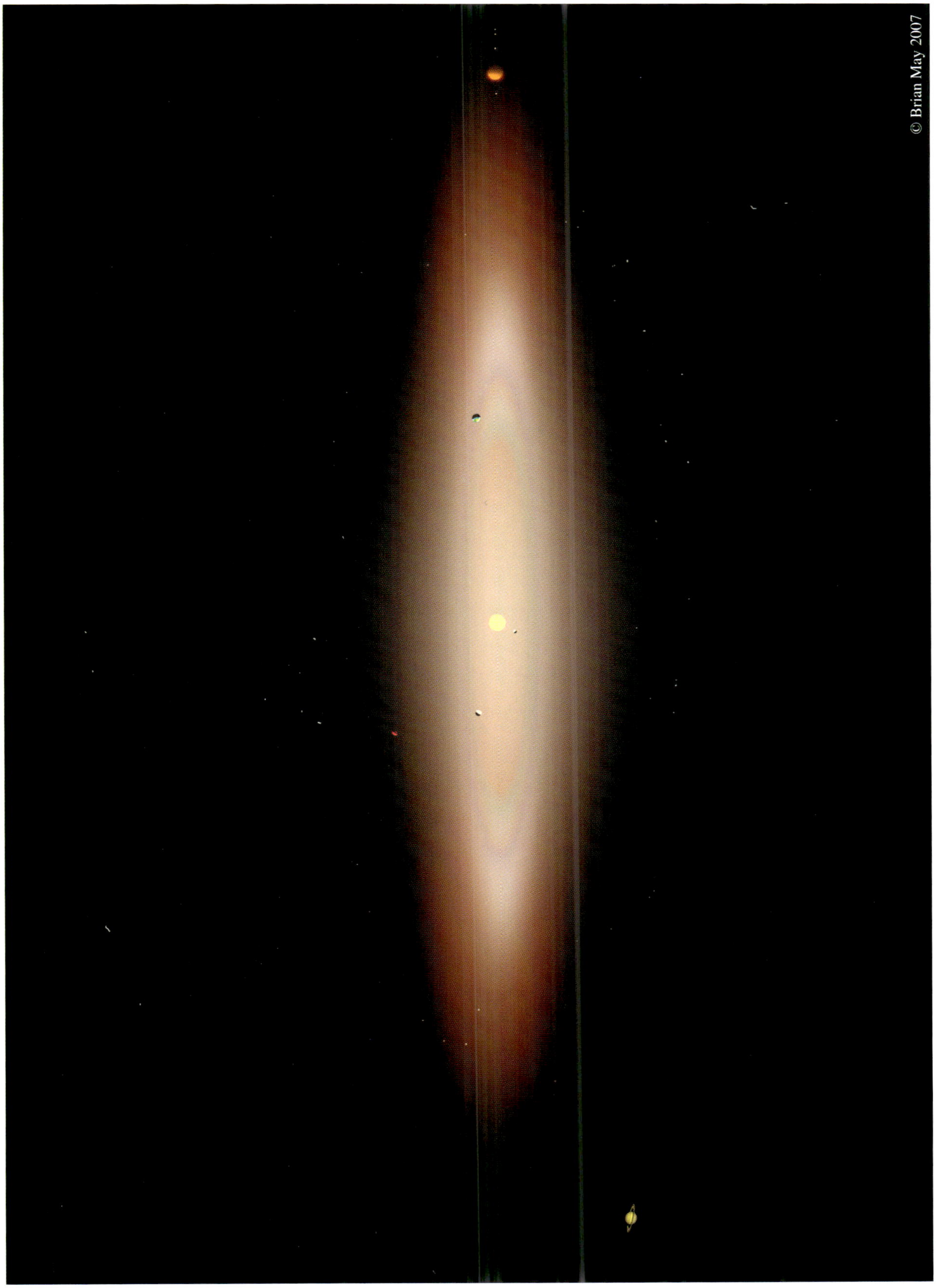
© Brian May 2007

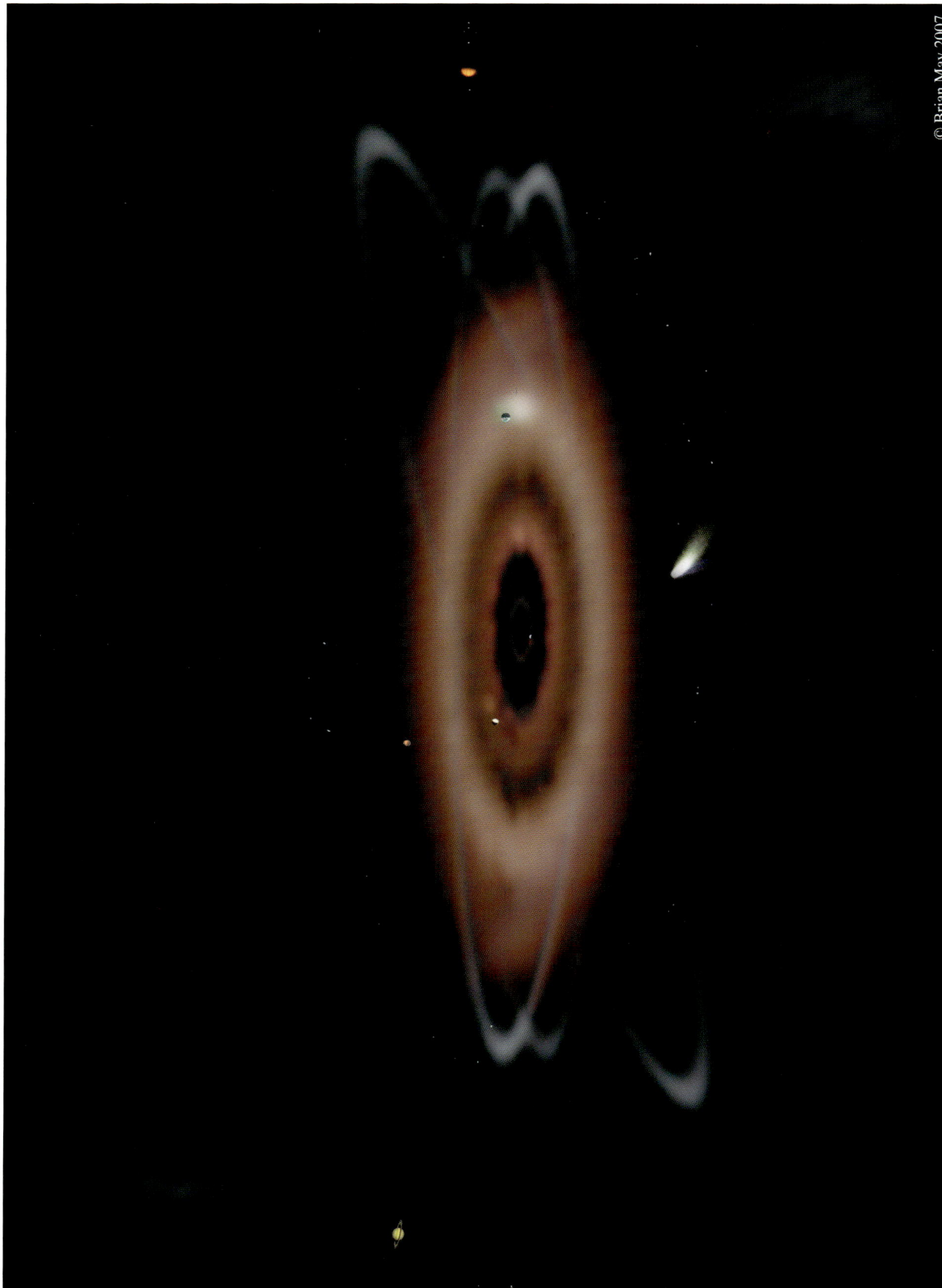
© Brian May 2007

© Brian May 2007

"In the sweat of thy face shalt thou eat bread, til thou return unto the ground;

for out of it wast thou taken:

for dust thou art,

and unto dust shalt thou return."

from the Bible, Genesis 3:19 (King James Version).

Appendix 1 Program ZLPLT

Program to process a night's raw data, subtracting the dark count, dividing out the premonochromating filter, dividing by the compensation channel count, and outputting the fully reduced scans on to cards.

Program ZLPLT page 1

```
      PROGRAM ZLPLT1(INPUT,OUTPUT,TAPE5=INPUT,TAPE6=OUTPUT,PUNCH)
      DIMENSION TITLE(76), PLOT(120)
      COMMON IDAT(50,16), SLMP(50,50), FINP(50), CFILT(50), NUM(50),
     1       SIG(50,50), COMP(50,50), FLMP(50), ZL(50), EZL(50), I,L
      DATA SET/3HSET/, DUMP/4HDUMP/, BLANK/1H /, STAR/1HX/
C     CLEAR ARRAYS AND READ ONE EVENINGS DATA INTO CORE.
 17   I=0
      READ(5,1003) (TITLE(K),K=1,76), SPLOD
      WRITE(6,1004) (TITLE(K),K=1,76), SPLOD
      DO 103 K1=1,50
      FINP(K1)=0.0
      CFILT(K1)=0.0
      ZL(K1)=0.0
      EZL(K1)=0.0
      NUM(K1)=0
      DO 108 K2=1,30
      SLMP(K1,K2)=0.0
      SIG(K1,K2)=0.0
      COMP(K1,K2)=0.0
108   FLMP(K2)=0.0
      DO 103 K3=1,16
103   IDAT(K1,K3)=0
 11   I=I+1
      READ(5,1000) (IDAT(I,J),J=1,16)
      IF(IDAT(I,1).EQ.0)        GO TO 10
 21   IF(IDAT(I,12)) 22,18,23
 22   N=-1*(IDAT(I,12))
      NUM(I)=N
      READ(5,1001) (SLMP(I,J),J=1,N)
      GO TO 24
 23   NUM(I)=(IDAT(I,14)-IDAT(I,12))/IDAT(I,13)+1
 24   K=NUM(I)+1
      IF(IDAT(I,2).EQ.1)        READ(5,1005) FINP(I)
      READ(5,1002) (SIG(I,J),COMP(I,J), J=1,K)
      GO TO 11
C     FIND FILTER SCAN AND DARK COUNT. CALCULATE MEAN DARK AND
C     CREATE LAMP ARRAYS.
 10   CONTINUE
      L=I-1
      IDK=0
      IFILT=0
      DO 100 K1=1,L
      IF(IDAT(K1,2).EQ.4)       IDK=K1
      IF(IDAT(K1,2).EQ.2)       IFILT=K1
      IF(IDAT(K1,12).LT.0)      GO TO 100
      SUM=FLOAT(IDAT(K1,12)-IDAT(K1,13))
      II=NUM(K1)
      DO 102 K2=1,II
      SUM=SUM+FLOAT(IDAT(K1,13))
102   SLMP(K1,K2)=SUM
100   CONTINUE
      IF(IDK.EQ.0.OR.IFILT.EQ.0)            GO TO 18
      NDK=NUM(IDK)
      SUM=0.0
      DSUM=0.0
      DO 101 K=1,NDK
      DSUM=DSUM+COMP(IDK,K)
101   SUM=SUM+SIG(IDK,K)
      DDK=DSUM/FLOAT(NUM(IDK))
      DARK=SUM/FLOAT(NUM(IDK))
C     CHECK CORRECT NUMBER OF LAMPS SEEN.
      J=0
 12   J=J+1
      L=I-1
      IF(J.EQ.I)                GO TO 16
      II=NUM(J)
      JJ=II+1
      IERR=0
      IF(SIG(J,II).EQ.0.0)      IERR=1
      IF(SIG(J,JJ).NE.0.0)      IERR=2
C     SHIFT FILTER SCAN TO COINCIDE WITH SIGNAL SCAN.
      CALL FIND(IFILT,FFPK,KAOS)
      CALL FIND(J,SFPK,KAOS)
      IF(KAOS.EQ.1)             GO TO 18
      PKDIF=SFPK-FFPK
      DO 106 K=1,50
      ZL(K)=0.0
106   EZL(K)=0.0
      Z=0.0
      ZM=0.0
      IF(IDAT(J,2).NE.3.OR.IERR.NE.0)            GO TO 15
      NFILT=NUM(IFILT)
```

```
        DO 105 L=1,II
        XX=SLMP(J,L)-PKDIF
        DO 104 M=1,NFILT
        K=NFILT+1-M
        FLMP(M)=SLMP(IFILT,M)
        IF(FLMP(M).LE.XX)          IX1=M
104     IF(FLMP(K).GT.XX)          IX2=K
        IF(FLMP(1).GE.XX)          IX1=1
        IF(FLMP(1).GE.XX)          IX2=2
        IF(FLMP(NFILT).LE.XX)      IX1=NFILT-1
        IF(FLMP(NFILT).LE.XX)      IX2=NFILT
        CFILT(L)=(SIG(IFILT,IX1)-SIG(IFILT,IX2))*(FLMP(IX2)-XX)/
   1             (FLMP(IX2)-FLMP(IX1))+SIG(IFILT,IX2)
C     CALCULATE CORRECTED Z.L. SCAN AND ERROR.
      DARK=PMOD*SMEAN
        ZL(L)=(SIG(J,L)-DARK)/((COMP(J,L)-DDK)*CFILT(L))
        Z=Z+ZL(L)
105     EZL(L)=SQRT(SIG(J,L)+DARK)/((COMP(J,L)-DDK)*CFILT(L))
        ZM=Z/FLOAT(II)
        DO 107 K=1,II
      DO 31 KKK=1,120
 31   PLOT(KKK)=BLANK
        ZL(K)=ZL(K)/ZM
      IX=IFIX(ZL(K)*60.0)
      DO 32 III=1,IX
 32   PLOT(III)=STAR
      WRITE(6,1013)(PLOT(KK),KK=1,120)
107     EZL(K)=EZL(K)/ZM
 15     WRITE(6,1007) (IDAT(J,K), K=1,16), IERR, SFPK, PKDIF
        IF(IDAT(J,2).NE.3.AND.SPLOD.NE.DUMP)         GO TO 19
        IF(IDAT(J,2).EQ.3)         WRITE(6,1011) ZM
        WRITE(6,1008) (SLMP(J,M), SIG(J,M), COMP(J,M),  ZL(M), EZL(M),
   1    M=1,II)
 19     CONTINUE
        GO TO 12
 18     WRITE(6,1010)
 16     READ(5,1009) TEST
        IF(TEST.EQ.SET)            GO TO 17


1000 FORMAT(16I5)
1001 FORMAT(20F4.1)
1002 FORMAT(12F6.1)
1003 FORMAT(76A1,A4)
1004 FORMAT(1H1,//1H ,76A1,A4)
1005 FORMAT(F5.1)
1007 FORMAT(1H0,3I4,1H-,2(I2,1H/),I4,3X,3HGMT,I2,1H.,I2,3X,2HST,I2,1H.,
    1I2,3X,1HT,I2,3X,I3,1H(,I2,1H),I3,3X,2HRA,I4,2X,3HDEC,I4,3X,2H**,
    2I1,2H**,3X,4HPLAK,E11.4,E14.4)
1008 FORMAT(1H ,5E20.6)
1009 FORMAT(A3)
1010 FORMAT(1H1,24(/),60X,4HTWIT,30(/),100X16H(C)  Z.P.I. 1972)
1011 FORMAT(22H NORMALIZATION FACTOR=,E12.6)
1012 FORMAT(3E20.6)


      SUBROUTINE FIND(J,FPK,KAOS)
      COMMON IDAT(50,16), SLMP(50,50), FINP(50), CFILT(50), NUM(50),
   1        SIG(50,50), COMP(50,50), FLMP(50), ZL(50), EZL(50), I,L
        KAOS=0
        IF1=0
        IF2=0
        DO 100 K=1,J
100     IF(IDAT(K,2).EQ.1)         IF1=K
        JJ=L-J+1
        DO 101 K=1,JJ
101     IF(IDAT((I-K),2).EQ.1)   IF2=I-K
        IF(IF1.EQ.0.AND.IF2.EQ.0)        GO TO 10
        IF(IF1.EQ.0)               FPK=FINP(IF2)
        IF(IF2.EQ.0)               FPK=FINP(IF1)
        IF(IF1.EQ.0.OR.IF2.EQ.0)         GO TO 11
        GMT1=FLOAT(IDAT(IF1,7))+(FLOAT(IDAT(IF1,8)))/60.0
        GMT2=FLOAT(IDAT(IF2,7))+(FLOAT(IDAT(IF2,8)))/60.0
        GMTX=FLOAT(IDAT(J,7))+(FLOAT(IDAT(J,8)))/60.0+((FLOAT(NUM(J)))
   1        /2.0)*(FLOAT(IDAT(J,11)+5))/3600.0
        X1=GMT1+12.0
        X2=GMT2+12.0
        X=GMTX+12.0
        IF(X1.GT.24.0)             X1=X1-24.0
        IF(X2.GT.24.0)             X2=X2-24.0
        IF(X .GT.24.0)             X=X-24.0
        Y1=FINP(IF1)
        Y2=FINP(IF2)
        IF(X1.NE.X2)               FPK=(Y1-Y2)*(X2-X)/(X2-X1)+Y2
        IF(X1.EQ.X2)               FPK=Y1
        GO TO 11
 10     KAOS=1
 11     RETURN
        END
```

Program ZLPLT page 2

Appendix 2 Programs which fit curves to the fully reduced data points

LSQFIT accepts the punched card fully reduced data and calculates a least-squares fitted polynomial curve of any specified order (to a maximum of six, in this case). Output is on cards, to be read as input by the programs in Appendix 2, which plot the curves and estimate the mean wavelength shift.

GAUSSN accepts the same data cards and calculates a least-squares fitted Gaussian curve, outputting parameters: mean wavelength shift, line depth and line width. Output is on cards, which are read by the program GAUSPT, a plotting program.

```
      PROGRAM LSQFIT(INPUT,OUTPUT,TAPE1=INPUT,TAPE2=OUTPUT,PUNCH,TAPE3=P
     1UNCH)
C     LEAST SQUARES FIT OF POLYNOMIAL, WITH STATISTICAL ANALYSIS
C
      DIMENSION X(200),Y(200),WT(200),RESID(200),U(200),V(200),
     1VAR(2000),FITPOL(21),F(21),ORTPOL(21),H(21),SSQS(21),C(21)
     2NOMAX(21),RESMAX(21),DINV(21),EL(210),COV(210)
C
      NOUT=2
      NIN=1
    1 READ(NIN,82)KELONG,SENSE
   82 FORMAT(I15,F15.1)
      READ(NIN,100) NUMBER,KDEG,AP,EQWT,WRITER
  100 FORMAT(2I15,3F15.1)
   86 FORMAT(I15,F15.1,2I15)
      WRITE(3,86)KELONG,SENSE,NUMBER,KDEG
      NDIM=2000
      KD=21
      KE=210
      IF(AP-0.5)4,30,30
    4 READ(NIN,101) FIRST X,DELTA X
  101 FORMAT(2F20.9)
      IF(EQ WT-0.5)23,22,22
   30 IF(EQ WT-0.5)21,20,20
C
   20 READ(NIN,102) (X(I),Y(I),WT(I),I=1,NUMBER)
  102 FORMAT(3F20.9)
      GO TO 3
   21 READ(NIN,103) (X(I),Y(I),I=1,NUMBER)
  103 FORMAT(4F20.9)
      DO 7 I=1,NUMBER
    7 WT(I)=1.0
      GO TO 3
C
   22 READ(NIN,104) (Y(I),WT(I),I=1,NUMBER)
  104 FORMAT(4F20.9)
      DO 8 I=1,NUMBER
      X(I)=FIRST X
    8 FIRST X=FIRST X + DELTA X
      GO TO 3
C
   23 READ(NIN,105) (Y(I),I=1,NUMBER)
  105 FORMAT(4F20.9)
      DO 9 I=1,NUMBER
      X(I)=FIRST X
      FIRST X=FIRST X + DELTA X
    9 WT(I)=1.0
C
C     NORMALISATION OF Y(I)
    3 S=0.0
      DO 190 I = 1,NUMBER
  190 S = S+Y(I)
      S = S/FLOAT(NUMBER)
      DO 191 I = 1,NUMBER
  191 Y(I) = Y(I)/S
C
C     WT() IS READ IN AS STANDARD DEVIATION, BUT MUST BE CONVERTED
C     TO RELATIVE WEIGHT  FOR USE IN PROGRAM
C
      S = 0.0
      DO 192 I = 1,NUMBER
      Z = 1.0/(WT(I)*WT(I))
      WT(I) = Z
  192 S=S+WT(I)
      S = S/FLOAT(NUMBER)
      DO 193 I = 1,NUMBER
      Z = WT(I)/S
  193 WT(I) = Z
C
      CALL POLYNL
     1(NUMBER,KDEG,WRITER,KD,NDIM,KE,X,Y,WT,RESID,U,V,VAR,FITPOL,
     2F,ORTPOL,H,SSQS,C,RESMAX,NOMAX,DINV,EL,COV,KELONG,SENSE)
      GO TO 1
      END


      SUBROUTINE POLYNL
     1 (NUMBER,KDEG,WRITER,KD,NDIM,KE,X,Y,WT,RESID,U,V,VAR,FITPOL,
     2F,ORTPOL,H,SSQS,C,RESMAX,NOMAX,DINV,EL,COV,KELONG,SENSE)
      DIMENSIONX(NDIM),Y(NDIM),WT(NDIM),RESID(NDIM),U(NDIM),V(NDIM),
     1VAR(NDIM),FITPOL(KD),F(KD),ORTPOL(KD),H(KD),SSQS(KD),C(KD),
     2RESMAX(KD),NOMAX(KD),DINV(KD),EL(KE),COV(KE)
C
C     =RESID(I)= HOLDS THE CURRENT DIFFERENCE BETWEEN THE DATA Y(I) AND
C     THE FITTED VALUE, AND IS ALTERED AFTER EACH INCREASE OF THE DEGREE
C     OF THE POLYNOMIAL. =U(I)= AND =V(I)= ARE USED IN THE RECURRENCE
C     RELATION TO CALCULATE THE VALUE OF AN ORTHOGONAL POLYNOMIAL IN
C     TERMS OF THE PREVIOUS TWO ORTHOGONAL POLYNOMIALS.
C     =VAR(I)= CONTAINS THE VARIANCE OF A FITTED Y VALUE, AND INCREASES
C     WITH EACH INCREASE OF DEGREE OF THE FITTED POLYNOMIAL.
      NOUT=2
      DO 1 I=1,NUMBER
      RESID(I)=Y(I)
      U(I)=1.0
      V(I)=0.0
    1 VAR(I)=0.0
C
```

Program LSQFIT page 1

Program LSQFIT page 2

```
C      IN ORDER TO AVOID ZERO SUBSCRIPTS, THE J=TH POLYNOMIAL IS REFERRED
C      TO BY THE INDEX (J+1) OR (J+2). =FIT POL= AND =ORT POL= REFER TO
C      THE FITTED AND THE ORTHOGONAL POLYNOMIALS RESPECTIVELY. =C(J+2)=
C      IS THE LEAST SQUARES ESTIMATE OF THE BEST MULTIPLE OF THE J=TH
C      ORTHOGONAL POLYNOMIAL.
       K2=KDEG + 3
       DO 22 J=1,K2
       FIT POL(J) =0.0
       F(J)=0.0
       ORT POL(J) = 0.0
   22  H(J)=0.0
C
       ORT POL(2) =1.
       C(1)=0.0
       WRITE(NOUT,96)KELONG
   96  FORMAT(14H1ELONGATION = ,I6,9H DEGREES )
       IF(SENSE-0.5)198,199,199
  198  WRITE(NOUT,98)
   98  FORMAT(15H EAST (EVENING))
       GO TO 400
  199  WRITE(NOUT,99)
   99  FORMAT(15H WEST (MORNING))
  400  WRITE(NOUT,101)
  101  FORMAT(62H0POWER      FITTED.          ORTHOGONAL       DEGREE     D
      1IVISOR /59H OF  X    POLYNOMIAL         POLYNOMIAL        J
      2  DJ)

       CALL POLFIT
      1(NUMBER,KDEG,WRITER,KD,NDIM,KE,X,Y,WT,RESID,U,V,VAR,FITPOL,F,
      2ORTPOL,H,SSQS,C,RESMAX,NOMAX,DINV,EL,COV)
C
       RETURN
        END


       SUBROUTINE POLFIT
      1(NUMBER,KDEG,WRITER,KD,NDIM,KE,X,Y,WT,RESID,U,V,VAR,FITPOL,F,
      2ORTPOL,H,SSQS,C,RESMAX,NOMAX,DINV,EL,COV)
       DIMENSIONX(NDIM),Y(NDIM),WT(NDIM),RESID(NDIM),U(NDTM),V(NDIM),
      1VAR(NDIM),FITPOL(KD),F(KD),ORTPOL(KD),H(KD),SSOS(KD),C(KD),
      2RESMAX(KD),NOMAX(KD),DINV(KD) ,EL(KE),COV(KE)
C
C      =DIVE IS THE DIVISOR USED TO CALCULATE VAR(C).
       NOUT=2
       DIV=1.0
       DINV(1)=0.0
       A=0.0
       B=-1.0
       KPLUS=KDEG+1
C
       DO 60 J=1,KPLUS
       EX=0.0
       EY=0.0
       Z=0.0
       BIG RES =0.0
       NO RES = 0
C
       DO 61 I=1,NUMBER
       IF(J-1)49,49,6
    6  VAR(I)=VAR(I)+V(I)*V(I)*DINV(J-1)
       RESID(I)=RESID(I)-C(J)*V(I)
    7  IF(WT(I)-0.0)49,49,64
   64  ABS RES = ABS (RESID(I))
       IF(ABS RES - BIG RES)49,49,62
   62  BIG RES = ABS RES
       NO RES = I
   49  W=(X(I)-A)*V(I) - B*U(I)
       U(I)=V(I)
       V(I)=W
       W=WT(I)*V(I)*V(I)
       EX=EX+W
       W=W*X(I)
       EY=EY+W
       W=RESID(I)*WT(I)*V(I)
   61  Z=Z+W
C
C
C      =RES MAX(J)= AND =NO MAX(J)= CONTAIN THE BIGGEST RESIDUAL AND ITS
C      NUMBER AFTER FITTING THE POLYNOMIAL OF DEGREE (J-1). =S SQS= IS
C      THE REDUCTION IN THE SUM OF SQUARES OF THE RESIDUALS.
    8  A=EY/EX
       B=EX/DIV
       DIV=EX
       RES MAX(J) = BIG RES
       NO MAX(J) = NO RES
       C(J+1)=Z/EX
       S SQS(J) = Z*C(J+1)
       J POWER = J-1
       WRITE(NOUT,101)
  101  FORMAT(1HJ)
C
       DO 66 I=1,J
       NT= I+1
```

```
      FIT POL(NT) = FIT POL(NT) + C(J+1)*ORT POL(NT)
      I POWER = I-1
      IF(I-J)9,65,65
    9 WRITE(NOUT,102)IPOWER,FIT POL(NT),ORT POL(NT)
  102 FORMAT(1H ,I3,2E19.8)
      GO TO 66
   65 WRITE(NOUT,103)IPOWER,FIT POL(NT),J POWER,DIV
  103 FORMAT(1H ,I3, E19.8,19H      0.10000000E 01,I6,E18.8)
   66 CONTINUE
      DINV(J)=1.0/DIV
      IF(WRITER-3.5)5,5,11
   11 IF(J-1)3,10,3
C
C     THE COEFFICIENTS OF THE ORTHOGONAL POLYNOMIALS ARE STORED IN THE
C     ARRAY =EL(210)=, FOR USE IN SUBROUTINE COVAR.
   10 EL(1)=1.0
      GO TO 5
    3 N1=(J*(J-1))/2
      DO 2 I=1,J
      L=N1+I
    2 EL(L)=ORTPOL(I+1)
    5 DO 63 I=1,NT
      NS=I+1
   63 H(NS) = ORT POL(I) - A*ORT POL(NS) - B*F(NS)
      DO 60 I=1,NT
      NS=I+1
      F(NS) = ORT POL(NS)
   60 ORT POL(NS) = H(NS)
      LIMIT = KPLUS +1
      WRITE(3,88)(FIT POL(I),I = 2,LIMIT)
   88 FORMAT(4E20.8)
C
      DET=DINV(KPLUS)
      CALL EST EY
     1(NUMBER,KDEG,WRITER,KPLUS,DET,KD,NDIM,KE,X,Y,WT,RESID,U,V,VAR,
     2FITPOL,F,ORTPOL,H, SSQS,C,RESMAX,NOMAX,DINV,EL,COV )
      RETURN
      END
```

Program LSQFIT page 3

```
      SUBROUTINE EST EY
     1(NUMBER,KDEG,WRITER,KPLUS,DET,KD,NDIM,KE,X,Y,WT,RESID,U,V,VAR,
     2FITPOL,F,ORTPOL,H, SSQS,C,RESMAX,NOMAX,DINV,EL,COV)
      DIMENSIONX(NDIM),Y(NDIM),WT(NDIM),RESID(NDIM),U(NDIM),V(NDIM),
     1VAR(NDIM),FITPOL(KD),F(KD),ORTPOL(KD),H(KD),SSQS(KD),C(KD),
     2RESMAX(KD),NOMAX(KD),DINV(KD),EL(KE),COV(KE)
C
      NOUT=2
      IF(WRITER - 0.5)43,43,56
   56 IF(WRITER - 1.5)42,42,57
   42 WRITE(NOUT,104)
  104 FORMAT(1H1,
     1///////51H   NO        X(I)              Y(I)            EST MEAN   )
      GO TO 44
   57 WRITE(NOUT,105)
  105 FORMAT(1H1,
     1///////51H   NO        X(I)              Y(I)            EST MEAN    ,
     15X, 9HRESIDUALS  )
   44 WRITE(NOUT,106)
  106 FORMAT(1H ,41X,8HOF Y(XI))
C
C     THE RESIDUAL SUM OF SQUARES IS STORED IN =RES SQS=, AND THE NUMBER
C     OF DEGREES OF FREEDOM IN =NDF=. =EST Y= IS THE FITTED Y VALUE.
   43 RES SQS = 0.0
      BIG RES = 0.0
      NDF = -KPLUS
      K2=KDEG+2
C
   14 DO 4 I=1,NUMBER
      RESID(I)=RESID(I) - V(I)*C(K2)
      VAR(I)=V(I)*V(I)*DINV(KPLUS)+VAR(I)
      EST Y = Y(I) - RESID(I)
C
      IF(WT(I)-0.0)50,50,59
   59 NDF=NDF+1
      ABS RES = ABS (RESID(I))
      IF(ABS RES - BIG RES)50,50,58
   58 BIG RES = ABS RES
      NO RES = I
   50 IF(WRITER-0.5)4,55,55
   55 IF(WT(I)-0.0)1,1,54
   54 IF(WRITER-1.5) 541,541,542
  541 WRITE(NOUT,107)I,X(I),Y(I),ESTY
  107 FORMAT(1H  ,I4,E15.4,2E16.6)
      GO TO 85
  542 WRITE(NOUT, 108)I,X(I),Y(I),ESTY,RESID(I)
  108 FORMAT(1H ,I4,E15.4,3E16.6)
      GO TO 85
```

Program
LSQFIT
page 4

```
    1 WRITE(NOUT,109)I,X(I),EST Y
  109 FORMAT(1H ,I4,E15.4,9X,2H**,5X,E16.6,11X,2H**)
   85 WRITE(3,84)X(I),Y(I),ESTY
   84 FORMAT(3E15.5)
    4 RES SQS = RES SQS + WT(I)*RESID(I)*RESID(I)
C
      RES MAX(K2) = BIG RES
      NO MAX(K2) = NO RES
C
      CALL STD DEV(NUMBER,KPLUS,WRITER,NDF,RES SQS,KD,NDIM, KE,
     1X,Y,WT,RESID,U,V,VAR, FITPOL,F,ORTPOL,H,SSQS,C,RESMAX,NOMAX,
     2DINV,EL,COV)
      RETURN
       END
```

```
      SUBROUTINE STD DEV (NUMBER,KPLUS,WRITER,NDF,RES SQS,KD, NDIM,KE,
     1X,Y,WT,RESID,U,V,VAR, FITPOL,F,CRTPOL,H,SSQS,C,RESMAX,NOMAX,
     2DINV,EL,COV)
      DIMENSIONX(NDIM),Y(NDIM),WT(NDIM),RESID(NDIM),U(NDIM),V(NDIM),
     1VAR(NDIM),FITPOL(KD),F(KD),ORTPOL(KD),H(KD),SSQS(KD),C(KD),
     2RESMAX(KD),NOMAX(KD),DINV(KD),EL(KE),COV(KE)
      NOUT=2
C     =AVE SQ= IS THE MEAN SQUARE RESIDUAL. =SD EY= CONTAINS TWICE THE
C     STANDARD DEVIATION OF =EST Y=. THE LATTER IS ESSENTIALLY THE
C     ESTIMATED MEAN OF A RANDOM VARIABLE Y(X). THE RANDOM VARIATION OF
C     Y(X) ABOUT ITS MEAN IS PREDICTED BY =SD Y=.
      IF(NDF-0)17,17,37
   37 AVE SQ = RES SQS/FLOAT (NDF)
   17 IF(WRITER-2.5)40,40,30
   30 WRITE(NOUT,110)
  110 FORMAT(1H1,
     1///////45H    NO  2(STD DEV OF EST     2(STD DEV OF Y(XI)/
     1       45H          MEAN OF Y(XI))        ABOUT EST MEAN)/1H )
      JUMP=11
C
    8 DO 38 I=1,NUMBER
      VAR(I) = AVE SQ * VAR(I)
      SD EY    = SCRT (VAR(I))*2.0
      SD Y    = SCRT (VAR(I) + AVE SQ) * 2.0
      IF(I-JUMP)38,39,39
   39 JUMP=JUMP+10
      WRITE(NOUT,116)
  116 FORMAT(1H )
   38 WRITE(NOUT,111)I,SD EY,SDY
  111 FORMAT(1H ,I5,E15.4,E21.4)
   40 WRITE(NOUT,112)
  112 FORMAT(1H1,
     1///////74H  DEG J    SUM OF SQUARES     F RATIO      PARAMS  CJ
     1MAX RESIDUAL   NO /1H )
      A = 1.0/AVE SQ
C
C     THE =F RATIO= PROVIDES A STATISTICAL MEASURE OF THE SIGNIFICANCE
C     OF THE GIVEN DEGREE OF THE POLYNOMIAL.
      DO 31 J=1,KPLUS
      J POWER = J-1
      J PLUS = J+1
      F RATIO = A * S SQS(J)
   31 WRITE(NOUT,113)JPOWER,S SQS(J),F RATIO, C(J PLUS),
     1 RES MAX(J PLUS), NO MAX(J PLUS)
  113 FORMAT(1H ,I4,E19.6,F12.2,E16.6,E16.6,I5)
C
      WRITE(NOUT,114)
  114 FORMAT(//22H D/F   RES SUM OF SOS     )
      IF(NDF-0)44,44,34
   34 WRITE(NOUT,115)
  115 FORMAT (1H+,25X,29HMEAN  SQUARE    ROOT M.S.          )
   44 WRITE(NOUT,117) NDF,RES SQS
  117 FORMAT(1H0,I4,E16.6)
      IF(NDF-0)42,42,32
   32 ROOT MS = SQRT (AVE SQ)
      WRITE(NOUT,118)AVESQ,ROOT MS
  118 FORMAT(1H+,19X,E17.6,E17.6)
```

Program GAUSSN Page 1

```
      PROGRAM GAUSSN(INPUT,OUTPUT,TAPE1=INPUT,TAPE2=OUTPUT,PUNCH,TAPE3=P
     1UNCH)
C
C     PROGRAM FITS GAUSSIANS TO DATA
C     INPUT FORMAT SAME AS FOR LSQFIT BUT KDEG,AP,WRITER ARE IGNORED
C
      DIMENSION X(200),Y(200),WT(200),W(200),Z(200),A(4,5),V(4),
     1E(200)
      NOUT=2
      NIN=1
    1 READ(NIN,82)KELONG,SENSE
   82 FORMAT(I15,F15.1)
      IF(FLOAT(KELONG)-0.0)10,10,9
    9 READ(NIN,100) NUMBER,KDEG,AP,EQWT,WRITER
  100 FORMAT(2I15,3F15.1)
   86 FORMAT(I15,F15.1,I15)
      WRITE(3,86)KELONG,SENSE,NUMBER
      READ(NIN,101) FIRST X,DELTA X
  101 FORMAT(2F20.9)
      DO 8 I = 1,NUMBER
      X(I) = FIRST X
    8 FIRST X = FIRST X+DELTA X
      IF(EQ WT-0.5)23,22,22
   22 READ(NIN,104) (Y(I),WT(I),I=1,NUMBER)
  104 FORMAT(4F20.9)
      GO TO 3
   23 READ(NIN,105) (Y(I),I=1,NUMBER)
  105 FORMAT(4F20.9)
      DO 4 I=1,NUMBER
    4 WT(I)=1.0
C
C     NORMALISATION OF Y(I)
C
    3 S=0.0
      DO 190 I = 1,NUMBER
  190 S = S+Y(I)
      S = S/FLOAT(NUMBER)
      DO 191 I = 1,NUMBER
  191 Y(I) = Y(I)/S
C
C     WT() IS READ IN AS STANDARD DEVIATION, BUT MUST BE CONVERTED
C     TO RELATIVE WEIGHT  FOR USE IN PROGRAM
C
      S = 0.0
      DO 192 I = 1,NUMBER
      Z = 1.0/(WT(I)*WT(I))
      WT(I) = Z
  192 S=S+WT(I)
      S = S/FLOAT(NUMBER)
      DO 193 I = 1,NUMBER
      Z = WT(I)/S
  193 W(I) = Z
C
      WRITE(NOUT,96)KELONG
   96 FORMAT(14H1ELONGATION = ,I6,9H DEGREES )
      IF(SENSE-0.5)198,199,199
  198 WRITE(NOUT,98)
   98 FORMAT(15H EAST (EVENING))
      GO TO 400
  199 WRITE(NOUT,99)
   99 FORMAT(15H WEST (MORNING))
C
C     INITIAL VALUES FOR VB01A ARE SET
C
  400 N = NUMBER
      V(1) = 1.0
      V(2) = 0.5
      V(3) = 0.3
      V(4) = 0.0
      M = 4
      IA = 4
      MAXIT = 20
      CALL VB01A(X,Y,W,Z,N,V,E,M,A,IA,MAXIT)
      WRITE(3,72)(V(I),I = 1,4)
   72 FORMAT(4E20.8)
      WRITE(3,50)(X(I),Y(I),Z(I),I = 1,N)
   50 FORMAT(3E15.5)
      GO TO 1
   10 STOP
      END
```

Program GAUSSN Page 2

```
                 SUBROUTINE VB01A (X,Y,W,Z,N,V,E,M,A,IA,MAXIT)              VB01A001
000015           DIMENSION X(1),Y(1),W(1),Z(1),V(1),E(1),A(IA,1)            VB01A002
000015           DIMENSION VV(20),F(20),DV(20)                              VB01A003
000015           MP1=M+1                                                    VB01A006
000016           M1=M-1                                                     VB01A007
000017           MARK=0                                                     VB01A008
000020           IT=0                                                       VB01A009
000021           NDF=N-M                                                    VB01A010
000022       706 DO 400 I=1,M                                               VB01A011
000030           VV(I)=V(I)                                                 VB01A012
000031       400 CONTINUE                                                   VB01A013
000035        11 S=0.0                                                      VB01A014
000036           DO 1 I=1,M                                                 VB01A015
000040           A(I,MP1)=0.0                                               VB01A016
000045           DO 2 J=1,I                                                 VB01A017
000054           A(I,J)=0.0                                                 VB01A018
000055         2 CONTINUE                                                   VB01A019
000062         1 CONTINUE                                                   VB01A020
000065           DO 3 K=1,N                                                 VB01A021
000066           CALL DERIV(X(K),V,F,Z(K))                                  VB01A022
000103       403 WS=Y(K)-Z(K)                                               VB01A023
000105           S=S+WS**2*W(K)                                             VB01A024
000110           WS=WS*W(K)                                                 VB01A025
000111           DO 4 I=1,M                                                 VB01A026
000113           A(I,MP1)=A(I,MP1)+F(I)*WS                                  VB01A027
000123           DO 5 J=1,I                                                 VB01A028
000140           A(I,J)=A(I,J)+F(I)*F(J)*W(K)                               VB01A029
000142         5 CONTINUE                                                   VB01A030
000147         4 CONTINUE                                                   VB01A031
000152         3 CONTINUE                                                   VB01A032
000154           DO 12 I=1,M1                                               VB01A033
000155           J1=I+1                                                     VB01A034
000156           DO 13 J=J1,M                                               VB01A035
000170           A(I,J)=A(J,I)                                              VB01A036
000171        13 CONTINUE                                                   VB01A037
000176        12 CONTINUE                                                   VB01A038
000201           CALL MA04A(A,M,IA)                                         VB01A039
000210       602 IF(MARK) 35,303,35                                         VB01A040
000211       303 WS1=S/(100.0* NDF)                                         VB01A041
000214           DO 16 I=1,M                                                VB01A042
000226           E(I)=WS1*A(I,I)                                            VB01A043
000227        16 CONTINUE                                                   VB01A044
000231           DO 116 I=1,M                                               VB01A045
000243           IF(A(I,MP1)**2-E(I))116,116,17                             VB01A046
000250       116 CONTINUE                                                   VB01A047
000255        35 VAR=S/NDF                                                  VB01A048
000260           DO 20 I=1,M                                                VB01A049
000261           DO 21 J=1,M                                                VB01A050
000271           A(I,J)=A(I,J)*VAR                                          VB01A051
000272        21 CONTINUE                                                   VB01A052
000274           E(I)=SQRT(A(I,I))                                          VB01A053
000310        20 CONTINUE                                                   VB01A054
000313           PRINT 122                                                  VB01A055
000316       122 FORMAT(1H1,       4X,1HI,10X,4HV(I),16X,12HS.D. OF V(I),//)  VB01A056
000322           DO 23 I=1,M                                                VB01A057
000323           PRINT             24,I, V(I),E(I)                          VB01A058
000351        23 CONTINUE                                                   VB01A059
000354        24 FORMAT(I5,2E20.8)                                          VB01A060

000354           CALL SA01A(S,NDF,PROB)                                     VB01A061
000356           PRINT             25,S,NDF,VAR,PROB                        VB01A062
000375        25 FORMAT(///,1X,24HRESIDUAL SUM OF SQUARES=,E16.8            VB01A063
                1,//1X,19HDEGREES OF FREEDOM=,I3,                           VB01A064
                1//,1X,15HVARIANCE RATIO=,E16.8,                            VB01A065
                1//,1X,12HPROBABILITY=,F8.4)                                VB01A066
000401           CALL OA01A(A,M,M,IA)                                       VB01A067
000404           PRINT 28                                                   VB01A068
000413        28 FORMAT(1H1,10X,1HX,15X,5HOBS Y,15X,8HFITTED Y,//)          VB01A069
000417           DO 29 K=1,N                                                VB01A070
000420           PRINT             30,X(K),Y(K),Z(K)                        VB01A071
000451        29 CONTINUE                                                   VB01A072
000453        30 FORMAT(3E20.8)                                             VB01A073
000453       504 RETURN                                                     VB01A074
000454        17 IF(IT-MAXIT)32,32,22                                       VB01A075
000457        22 PRINT 34,IT                                                VB01A076
000465        34 FORMAT(22H TERMINATED, MORE THAN,I3,11H ITERATIONS,//)     VB01A077
000471           GO TO 35                                                   VB01A078
000471        32 IT=IT+1                                                    VB01A079
000473           DO 31 I=1,M                                                VB01A080
000504           VV(I)=V(I)                                                 VB01A081
000505           V(I)=VV(I)+A(I,MP1)                                        VB01A082
000507           DV(I)=A(I,MP1)                                             VB01A083
000507        31 CONTINUE                                                   VB01A084
000510           PRINT             40,IT,S,(VV(I),V(I),I=1,M)               VB01A085
000544        40 FORMAT(10H ITERATION,I3,2X,2HS=,E16.8,/,(2E20.8))          VB01A086
000550           S1=S                                                       VB01A087
000551           GO TO 11                                                   VB01A088
000552           END                                                        VB01A089
```

```
                 SUBROUTINE DERIV(X,V,F,FUNC)
000006           DIMENSION X(200),V(4),F(4)
          C
          C      THE FUNCTION IS DEFINED  Y = V1-V2EXP(-V3(X-V4)**2)
          C
000006           P = (X-V(4))**2
000007           S = -V(3)*P
000012           Q = EXP(S)
000017           FUNC = V(1)-V(2)*Q
000021           F(1) = 1.0
000023           F(2) = -Q
000024           F(3) = V(2)*P*Q
000026           F(4) = 2.0*V(2)*V(3)*(X-V(4))*Q
000032           F(4) = -F(4)
000033           RETURN
000033           END

                 SUBROUTINE MA04A(A,M,IA)
000005           DIMENSION A(IA,1),C(100),IND(100)
000005           MP1=M+1
000006       100 AMAX=0.0
000007           DO 2 I=1,M
000020           IND(I)=I
000021           IF(ABS(A(I,1))-AMAX)2,2,3
000024         3 AMAX=ABS(A(I,1))
000025           I4=I
000026         2 CONTINUE
000027           MM=M-1
000033           DO 111 J=1,MM
000034           IF(I4-J)6,6,4
000041         4 ISTO=IND(J)
000041           IND(J)=IND(I4)
000042           IND(I4)=ISTO
000043           DO 5 K=1,MP1
000055           STO=A(I4,K)
000056           A(I4,K)=A(J,K)
000057           A(J,K)=STO
000060         5 CONTINUE
000063         6 AMAX=0.0
000064           J1=J+1
000066           DO 11 I=J1,M
000067           A(I,J)=A(I,J)/A(J,J)
000075           DO 10 K=J1,MP1
000112           A(I,K)=A(I,K)-A(I,J)*A(J,K)
000114           IF  (K-J1)14,14,10
000116        14 IF(ABS(A(I,K))-AMAX)10,10,17
000121        17 AMAX=ABS(A(I,K))
000122           I4=I
000124        10 CONTINUE
000130        11 CONTINUE
000132       111 CONTINUE
000134        19 DO 127 I1=1,M
000136           I=M+1-I1
000137           IF(M-I)327,327,28
000141        28 I2=I+1
000143           DO 32 K=I2,M
000157           A(I,MP1)=A(I,MP1)-A(I,K)*A(K,MP1)
000161        32 CONTINUE
000165       327 A(I,MP1)=A(I,MP1)/A(I,I)
000174       127 CONTINUE
000177        65 DO 140 I1=1,MM
000201           I=M+1-I1
000202           I2=I-1
000204           DO 41 J1=1,I2
000205           J=I2+1-J1
000207           J2=J+1
000207           W1=-A(I,J)
000213           IF(I2-J2)141,43,43
000216        43 DO 42 K=J2,I2
000227           W1=W1-A(K,J)*C(K)
000231        42 CONTINUE
000234       141 C(J)=W1
000236        41 CONTINUE
000241           DO 40 K=1,I2
000251           A(I,K)=C(K)
```

Program
GAUSSN
Page 3

Program GAUSSN Page 4

```
000252       40 CONTINUE                                                       MA04A059
000255      140 CONTINUE                                                       MA04A060
000260          DO 150 I1=1,M                                                  MA04A061
000261          I=M+1-I1                                                       MA04A062
000262          I2=I+1                                                         MA04A063
000263          W=A(I,I)                                                       MA04A064
000270          DO 56 J=1,M                                                    MA04A065
000271          IF (I-J)52,53,54                                               MA04A066
000273       52 W1=0.0                                                         MA04A067
000274          GO TO 55                                                       MA04A068
000275       53 W1=1.0                                                         MA04A069
000277          GO TO 55                                                       MA04A070
000277       54 W1=A(I,J)                                                      MA04A071
000304       55 IF(I1-1)156,156,57                                             MA04A072
000307       57 DO 58 K=I2,M                                                   MA04A073
000321          W1=W1-A(I,K)*A(K,J)                                            MA04A074
000323       58 CONTINUE                                                       MA04A075
000327      156 C(J)=W1                                                        MA04A076
000331       56 CONTINUE                                                       MA04A077
000333          DO 50 J=1,M                                                    MA04A078
000346          A(I,J)=C(J)/W                                                  MA04A079
000347       50 CONTINUE                                                       MA04A080
000352      150 CONTINUE                                                       MA04A081
000355          DO 60 I=1,M                                                    MA04A082
000356       63 IF(IND(I)-I)61,60,61                                           MA04A083
000360       61 J=IND(I)                                                       MA04A084
000362          DO 62 K=1,M                                                    MA04A085
000374          STO=A(K,I)                                                     MA04A086
000375          A(K,I)=A(K,J)                                                  MA04A087
000376          A(K,J)=STO                                                     MA04A088
000376       62 CONTINUE                                                       MA04A089
000401          ISTO=IND(J)                                                    MA04A090
000402          IND(J)=J                                                       MA04A091
000403          IND(I)=ISTO                                                    MA04A092
000405          GO TO 63                                                       MA04A093
000407       60 CONTINUE                                                       MA04A094
000412       64 RETURN                                                         MA04A095
000413          END                                                            MA04A096

                SUBROUTINE  SA01A(X,N,PROBX2)                                  SA01A001
000005          DIMENSION C(16)                                                ICST0002
000005          DATA C     /1.47242543E-15,2.4603754E-14,3.9554295E-13,        ICST0A02
               1 5.94779401E-12,8.3507028E-11,1.08922104E-9,1.3122533E-8,      SA01A003
               2 1.45038522E-7,1.4589169E-6,1.32275132E-5,1.06837607E-4,       SA01A004
               3 7.57575758E-4,4.62962963E-3,2.38095238E-2,1.0E-1,0.0/,        ICST0005
               4 RRTP/.3989422804/                                             SA01A006
000005          IF (X-1.3)10,10,11                                             SA01A007
000007       10 P=.5                                                           SA01A008
000011          GO TO 30                                                       SA01A009
000011       11 P=0.                                                           SA01A010
000012          IF (N-1)30,40,41                                               SA01A011
000014       41 EN=N                                                           SA01A012
000015          IF (.875*X+1.0-EN)10,12,12                                     SA01A013
000021       12 IF(N-20)16,16,15                                               SA01A014
000023       16 XB=EXP(-.5*X)                                                  SA01A015
000026          M=N/2                                                          SA01A016
000032          DO 19 I=1,M                                                    SA01A017
000040          P=P*X/EN+1.0                                                   SA01A018
000042          EN=EN-2.0                                                      SA01A019
000043       19 CONTINUE                                                       SA01A020
000045          P=P*XB                                                         SA01A021
000047          IF (N-2*M)30,30,17                                             SA01A022
000053       17 P=2.*P*RRTP*SQRT(X)                                            SA01A023
000061       40 IF(X-10.0)21,21,30                                             SA01A024
000064       21 XD=-0.5*X                                                      SA01A025
000065          XC=RRTP*SQRT(X)                                                SA01A026
000070          C(16)=3.333333333 E-1                                          SA01A027
000074       22 Q=8.0117266E-17                                                SA01A028
000076          DO 23 I=1,16                                                   SA01A029
000103          Q=Q*XD+C(I)                                                    SA01A030
000105       23 CONTINUE                                                       SA01A031
000106          Q=(1.+Q*XD)*XC                                                 SA01A032
000111          IF (N-20)25,25,24                                              SA01A033
000115       24 P=.5-Q                                                         SA01A034
000117          GO TO 30                                                       SA01A035
000120       25 P=P+1.-2.*Q                                                    SA01A036
000124          GO TO 30                                                       SA01A037
000124       15 C(16)=3.333333333 E-1                                          SA01A038
000125          XC=4.5*EN                                                      SA01A039
000127          XE=(X/EN)**C(16)-1.0+1.0/XC                                    SA01A040
000135          XD=-.5*XE*XE*XC                                                SA01A041
000137          XC=XE*RRTP*SQRT(XC)                                            SA01A042
000143          IF (XD+5.)14,22,22                                             SA01A043
000150       14 P=0.0                                                          SA01A044
000151       30 PROBX2=P                                                       SA01A045
000152          RETURN                                                         SA01A046
000153          END                                                            SA01A047
```

Program GAUSSN Page 5

```
                 SUBROUTINE OA01A(A,M,N,IA)
000006           DIMENSION A(1)
000006           IF(F)21,20,21
000007      21   IM=0
000010           F=0
000011      20   IM=IM+1
000013           IP=1
000013           J3=(N-1)/5+1
000020           DO 101 J=1,J3
000021           J1=5*J-4
000023           J2=5*J
000024           IF(J2-N)6,6,7
000026       7   J2=N
000027      14   FORMAT(1H1,7HMATRIX ,I3,4X,4HPAGE,I3,///)
000027       6   PRINT 14,IM,IP
000037           IP=IP+1
000041           PRINT4,(J4,J4=J1,J2)
000055       4   FORMAT (I19,4I20)
000060           PRINT 3
000063       3   FORMAT (1X)
000066           M1=1
000066           M2=1
000067           M3=0
000071           DO 1 I=1,M
000072           K1=IA*(J1-1)+I
000075           K2=IA*(J2-1)+I
000100           PRINT 5,I,(A(K),K=K1,K2,IA)
000122       5   FORMAT (I5,5E20.8)
000125           IF (M1-5)9,10,9
000127      10   PRINT 8
000133       8   FORMAT (1X)
000136           M1=0
000137       9   IF (M2-30)11,12,11
000141      12   IF (M2-M) 13,1,1
000143      13   PRINT 14,IM,IP
000153           PRINT 4,(J4,J4=J1,J2)
000167           IP=IP+1
000171           M2=0
000174      11   M1=M1+1
000176           M2=M2+1
000177       1   CONTINUE
000202     101   CONTINUE
000204           RETURN
000205           END
```

Appendix 3 Programs which draw the fitted curves and extract parameters from the fits

POLYPT accepts the output cards from LSQFIT as input, for any order polynomial, plots the experimental points (x_i, y_i) and draws the curve of the polynomial, by calculating its value at $1/20^{th}$ angstrom intervals from the coefficients.

FLEXPT accepts the output cards of LSQFIT, for 4th order only, and calculates the mid-point of the line joining the inflexion points of the curve, giving an estimate of the mean wavelength shift.

MINIPT and MINIPS accept cards from LSQFIT 4th and 6th order polynomial respectively. They find the minimum point of the curve by successive approximation, giving a different estimate of the wavelength shift.

GAUSPT accepts the output of GAUSSN and draws the fitted Gaussian shape.

All exit on a blank card.

Program POLYPT

```
      PROGRAM POLYPT(INPUT,OUTPUT,TAPE1=INPUT,TAPE2=OUTPUT,TAPE 27)
C
C     CALCOMP PLOT OF POLYNOMIAL FITTED TO Z.L. AVERAGED SPECTRUM
C
      DIMENSION X(200),Y(200),YFIT(200),FIT POL(200)
    7 CALL START
  100 READ(1,1)KELONG,SENSE,NUMBEN,KDEG
    1 FORMAT(I15,F15.1,2I15)
      IF (FLOAT(KELONG)-0.0)10,10,9
C
C     COEFFICIENT OF THE I,TH POWER IN THE FITTED POLYNOMIAL
C     IS STORED IN FIT POL(I+1)
C
    9 KPLUS=KDEG+1
      READ(1,2)(FIT POL(I),I=1,KPLUS)
    2 FORMAT(4E20.8)
      READ(1,3)(X(I),Y(I),YFIT(I),I=1,NUMBEN)
    3 FORMAT(3E15.5)
C
C     FITTED POLYNOMIAL IS FOUND ANALYTICALLY FOR VALUES OF
C     X IN 1/20 ANGSTROM STEPS IN THE RANGE OF DATA
C
      CALL PLOT (4.0,0.0,-3)
      P = YFIT(1)/0.2-2.0
      CALL PLOT (X(1),P,3)
      Z = X(1)
   15 Z = Z+0.05
      IF (Z-X(NUMBEN))11,11,12
   11 ANA  = FIT POL(1)
      DO 13 I = 1,KDEG
      ZN = Z**I
   13 ANA  = ANA +FIT POL(I+1)*ZN
      ANA = ANA/0.2-2.0
      ANAL = ANA
      CALL PLOT (Z,ANAL,2)
      GO TO 15
   12 CALL PLOT(-4.0,0.0,-3)
C
      CALL SYMBOL(5.4,1.0,0.14,7HELONG.=,0.0,7)
      ELONG = FLOAT(KELONG)
      CALL NUMBER (6.4,1.0,0.14,ELONG,0.0,-1)
      IF(SENSE-0.5)4,5,5
    4 CALL SYMBOL(7.4,1.0,0.14,4HEAST,0.0,4)
      GO TO 6
    5 CALL SYMBOL(7.4,1.0,0.14,4HWEST,0.0,4)
    6 CALL AXIS(0.0,0.0,31HRELATIVE WAVELENGTH (ANGSTROMS),
     1-31,8.0,0.0,-4.0,1.0)
      CALL AXIS(4.0,0.0,31H                       INTENSITY,
     131,8.0,90.0,0.4,0.2)
      X(NUMBEN+1) = -4.0
      X(NUMBEN+2) = 1.0
      Y(NUMBEN+1) =0.4
      Y(NUMBEN+2) =0.2
      CALL LINE(X,Y,NUMBEN,1,-1,3)
    8 CALL PLOT (10.0,0.0,-3)
      GO TO 100
   10 CALL ENPLOT(10.0)
      END
```

Program FLEXPT

```
      PROGRAM FLEXPT(INPUT,OUTPUT,TAPE1=INPUT,TAPE2=OUTPUT)
C
      DIMENSION X(200),Y(200),YFIT(200),FIT POL(200)
      WRITE(2,23)
   23 FORMAT(21H0   ELONGATION    SHIFT)
  100 READ(1,1)KELONG,SENSE,NUMBEN,KDEG
    1 FORMAT(I15,F15.1,2I15)
      IF (FLOAT(KELONG)-0.0)10,10,9
    9 KPLUS=KDEG+1
      READ(1,2)(FIT POL(I),I=1,KPLUS)
    2 FORMAT(4E20.8)
      READ(1,3)(X(I),Y(I),YFIT(I),I=1,NUMBEN)
    3 FORMAT(3E15.5)
      SHIFT = -FIT POL(4)/FIT POL(5)
      SHIFT = SHIFT/4.0
   22 FORMAT(I10,F10.3)
      WRITE(2,22)KELONG,SHIFT
      GO TO 100
   10 STOP
      END
```

Program MINIPT

```
      PROGRAM MINIPT(INPUT,OUTPUT,TAPE1=INPUT,TAPE2=OUTPUT)
      DIMENSION X(200),Y(200),YFIT(200),FIT POL(200)
      WRITE(2,23)
   23 FORMAT(36H0  ELONGATION   SHIFT           LOOPS)
  100 READ(1,1)KELONG,SENSE,NUMBEN,KDEG
    1 FORMAT(I15,F15.1,2I15)
      IF  (FLOAT(KELONG)-0.0)10,10,9
    9 KPLUS=KDEG+1
      READ(1,2)(FIT POL(I),I=1,KPLUS)
    2 FORMAT(4E20.8)
      READ(1,3)(X(I),Y(I),YFIT(I),I=1,NUMBEN)
    3 FORMAT(3F15.5)
      I = 0
      ROOT = 0.0
    5 I = I+1
      RONE = ROOT
      RTWO =RONE*RONE
      RTHREE = RTWO*RONE
      DF = FIT POL(2)+2.0*FIT POL(3)*RONE+3.0*FIT POL(4)*RTWO
     1+4.0*FIT POL(5)*RTHREE
      DDF = 2.0*FIT POL(3)+6.0*FIT POL(4)*RONE+12.0*FIT POL(5)*RTWO
      DELTA = DF/DDF
      ROOT = ROOT-DELTA
      IF(DELTA-0.001)12,5,5
   12 WRITE(2,30)KELONG,ROOT,I
   30 FORMAT(I10,F10.3,I15)
      GO TO 100
   10 STOP
      END
```

Program MINIPS

```
      PROGRAM MINIPS(INPUT,OUTPUT,TAPE1=INPUT,TAPE2=OUTPUT)
C
C     SIXTH ORDER VERSION OF MINIPT
C
      DIMENSION X(200),Y(200),YFIT(200),FIT POL(200)
      WRITE(2,23)
   23 FORMAT(36H0  ELONGATION   SHIFT           LOOPS)
  100 READ(1,1)KELONG,SENSE,NUMBEN,KDEG
    1 FORMAT(I15,F15.1,2I15)
      IF  (FLOAT(KELONG)-0.0)10,10,9
    9 KPLUS=KDEG+1
      READ(1,2)(FIT POL(I),I=1,KPLUS)
    2 FORMAT(4E20.8)
      READ(1,3)(X(I),Y(I),YFIT(I),I=1,NUMBEN)
    3 FORMAT(3E15.5)
      I = 0
      ROOT = 0.0
    5 I = I+1
      RONE = ROOT
      RTWO = RONE*RONE
      RTHREE = RTWO*RONE
      RFOUR = RTHREE*RONE
      RFIVE =RFOUR*RONE
      DF = FIT POL(2)+2.0*FIT POL(3)*RONE+3.0*FIT POL(4)*RTWO
     1+4.0*FIT POL(5)*RTHREE
     2+5.0*FIT POL(6)*RFOUR+6.0*FIT POL(7)*RFIVE
      DDF = 2.0*FIT POL(3)+6.0*FIT POL(4)*RONE+12.0*FIT POL(5)*RTWO
     1+20.0*FIT POL(6)*RTHREE+30.0*FIT POL(7)*RFOUR
      DELTA = DF/DDF
      ROOT = ROOT-DELTA
      IF(DELTA-0.001)12,5,5
   12 WRITE(2,30)KELONG,ROOT,I
   30 FORMAT(I10,F10.3,I15)
      GO TO 100
   10 STOP
      END
```

```
      PROGRAM GAUSPT(INPUT,OUTPUT,TAPE1=INPUT,TAPE2=OUTPUT,TAPE27)
C
C     CALCOMP PLOT OF  GAUSSIAN  FITTED TO Z.L. AVERAGED SPECTRUM
C
      DIMENSION X(200),Y(200),YFIT(200),V(4),Z(100)
    7 CALL START
      B = 0.0
  100 READ(1,1)KELONG,SENSE,NUMBEN
    1 FORMAT(I15,F15.1, I15)
      IF (FLOAT(KELONG)-0.0)10,10,9
    9 READ(1,72)(V(I), I = 1,4)
   72 FORMAT(4E20.8)
      READ(1,50)(X(I),Y(I),Z(I),I = 1,NUMBEN)
   50 FORMAT(3E15.5)
C
C     FITTED  GAUSSIAN  IS FOUND ANALYTICALLY FOR VALUES OF
C     X IN 1/20 ANGSTROM STEPS IN THE RANGE OF DATA
C
      CALL PLOT (4.0,0.0,-3)
      Z = X(1)
      P = (Z-V(4))**2
      S = -V(3)*P
      Q = EXP(S)
      FUNC = V(1)-V(2)*Q
      ANAL = FUNC/0.2-2.0
      CALL PLOT(Z,ANAL,3)
   15 CONTINUE
      Z = Z+0.05
      IF (Z-X(NUMBEN))11,11,12
   11 P = (Z-V(4))**2
      S = -V(3)*P
      Q = EXP(S)
      FUNC = V(1)-V(2)*Q
      ANAL = FUNC/0.2-2.0
  114 CALL PLOT(Z,ANAL,2)
      GO TO 15
   12 CALL PLOT(-4.0,0.0,-3)
C
      CALL SYMBOL(5.4,1.0,0.14,7HELONG.=,0.0,7)
      ELONG = FLOAT(KELONG)
      CALL NUMBER (6.4,1.0,0.14,ELONG,0.0,-1)
      IF(SENSE-0.5)4,5,5
    4 CALL SYMBOL(7.4,1.0,0.14,4HEAST,0.0,4)
      GO TO 6
    5 CALL SYMBOL(7.4,1.0,0.14,4HWEST,0.0,4)
    6 CALL AXIS(0.0,0.0,31HRELATIVE WAVELENGTH (ANGSTROMS),
     1-31,8.0,0.0,-4.0,1.0)
      CALL AXIS(4.0,0.0,31H                      INTENSITY,
     131,6.0,90.0,0.4,0.2)
      X(NUMBEN+1) = -4.0
      X(NUMBEN+2) = 1.0
      Y(NUMBEN+1) =0.4
      Y(NUMBEN+2) =0.2
      CALL LINE(X,Y,NUMBEN,1,-1,3)
      IF(B-2.5)82,82,83
   82 CALL PLOT(0.0,7.0,-3)
      GO TO 8
   83 CALL PLOT(10.0,-21.0,-3)

      B = 0.0
      GO TO 100
    8 B = B+1.0
      GO TO 100
   10 CALL ENPLOT(10.0)
      END
```

Program
Gauspt

Appendix 4 STREAM and CONVL

Program STREAM Calculating Doppler shifts from a uniform dust flow model

For details see section 4.11.1.

Program STREAM

```
      PROGRAM STREAM(INPUT,OUTPUT,TAPE1=INPUT,TAPE2=OUTPUT,PUNCH,TAPE3=P
     1UNCH)
C
C     PROGRAM COMPUTES MEAN DOPPLER SHIFT FOR DUST FLOWING
C     LINEARLY THROUGH SOLAR SYSTEM
C
      DIMENSION Y(200)
      NOUT=2
      NIN=1
      READ(NIN,1)SLONG,VDUST,VEARTH
    1 FORMAT(F4.0,F4.1,F4.1)
      WRITE(NOUT,20)SLONG
   20 FORMAT(30H1LINEAR FLOW MODEL WITH SUN AT,F4.0,
     126HDEGREES ECLIPTIC LONGITUDE//)
      WRITE(NOUT,21)
   21 FORMAT(22H ELONG   SHIFT   STEPS/15H (DEG)   (ANGST)//)
      ELONG = 10.0
  100 X = 0.0
      J = 1
      TOP = 0.0
      BOT = 0.0
      THETAD = 271.0-SLONG-ELONG
    4 IF(THETAD-0.0)5,5,6
    5 THETAD = THETAD+360.0
    6 E = 3.14159*ELONG/180.0
      CE = COS(E)
      S = SIN(E)
   15 C = X-CE
      THETA = 180.0*ATAN(S/C)/3.14159
      IF (C -0.0)7,8,8
    7 THETA = THETA+180.0
    8 CONTINUE
      IF (THETA-0.0)25,26,26
   25 THETA = THETA+360.0
   26 CONTINUE
      THETAS = THETA-THETAD
      IF(THETAS-0.0)9,10,10
    9 THETAS = THETAS+360.0
   10 THETAS = 3.142*THETAS/180.0
      RHETAD = 3.142*THETAD/180.0
      DLAMDA = 5183.6*(VEARTH*SIN(E)     -VDUST*COS(RHETAD)
     1-VDUST*COS(THETAS))/(3.0*10.0**5)
      B = 1.0/(X**2+1.0-2.0*X*CE)
      T = DLAMDA*B
      IF(X-0.01)11,11,12
C     HALF VALUES ARE FOR THE FIRST TRAPEZIUM. THE LAST IS IGNORED
   11 B = B/2.0
      T = T/2.0
   12 TOP = TOP+T
      BOT = BOT+B
      RATIO = B/BOT
      IF(RATIO-0.01)13,13,14
C     SERIES TRUNCATED IF CONTRIBUTION IS LESS THAN 1 PER CENT.
   14 X = X+0.1
      J = J+1
      GO TO 15
   13 SHIFT = TOP/BOT
      WRITE(NOUT,16)ELONG,SHIFT.J
   16 FORMAT(1X,F4.0,3H    ,F6.3,3H    ,I3)
      WRITE(3,22)ELONG,SHIFT
   22 FORMAT(F4.0,F6.3)
      IF(ELONG-350.0)17,18,18
   18 GO TO 200
   17 ELONG = ELONG+10.0
      GO TO 100
  200 STOP
      END
```

Program CONVL Simulation of signal output of scanning Fabry-Perot spectrometer

The program reads into an array A(2001-4001) a spectrum at relatively high resolution, at 1/50th angstrom intervals. It then places in an array B(2001-4001) values of the interference filter transmission over the corresponding wavelength range. The two arrays are now multiplied together point by point, to place in the array E(2001-4001) the spectrum as modified by passage through the interference filter. Using the Airy function with values of finesse and etalon gap read from the data, a Fabry-Perot profile centred on the point E(3001) is put into the locations C(1-6001). The program then convolves arrays E and C, the sums of products for E(2001-4001) being used for shifts of the array C from –200 to +200, the sum for each shift giving one value in the array SUM(1-4001). Finally the result is divided by the filter shape as widened by convolution with the instrumental function, imitating the method used in the reduction of real scans.

At each stage a crude graphical output is printed; the process can be repeated for different etalon parameters.

DATA:

i) Spectrum – format as desired matched to a suitable read statement.

ii) Format (F2.0) FINESSE.

iii) Format (F3.0) ETALON GAP.

Blank card to exit.

Program CONVL

```
      PROGRAM CONVL ( INPUT, OUTPUT, TAPE5=INPUT, TAPE6=OUTPUT )
C
C     SCAN SIMULATION WITH AN IDEAL FABRY-PEROT          BHM 5
C
      DIMENSION A(6001),B(6001),C(6001),SUM(4001),E(6001)
      DO 1 I = 1,6001
      A(I) = 0.0
      B(I) = 0.0
    1 C(I) = 0.0
C
C     GENERATE SPECTRUM
      DO 995 I = 1,6001
  995 A(I) = 100.0
C
C     READ ELLIOTT PROGRAM PROFILE, ALPHA = 1, EPSILON =.30 DEG.
C
      READ(5,99)(A(K),K = 2899,3099,2)
   99 FORMAT(10F5.1)
      DO 88 K = 2899,3097,2
   88 A(K+1) =  (A(K)+A(K+2))/2.0
C
C     READ 6-ANGSTROM FILTER
      U = 2.77260/((6.0*50.0)**2)
      DO 67 K = 2001,4001,1
      X = FLOAT(K-3001)
   67 B(K) = 100.0*EXP(-U*X*X)
C     MULTIPLY LINE SHAPE AND FILTER TOGETHER
      DO 300 I = 1,6001
  300 E(I) = A(I)*B(I)
      CALL LGRAPH(E(2001),2001,1,10)
C
C     FABRY PEROT PROFILES                                    C( )
  777 FORMAT (F 2.0)
  774 READ (5,777) F
      IF (F .EQ. 0.0) GO TO 25
  776 FORMAT (F 3.0)
      READ(5,776)T
      P = 3.14159162
      W = 0.51836
      A = 2.0*T/W
      A = AINT (A)
      T = A*W/2.0
      FREE = W*W*10000.0/(2.0*T)
      R = FREE/F
      DO 775 I = 3001,6001
      S = SIN (P*2.0*T/W)
      C(I) = 1.0/(1.0+4.0*F*F*S*S/(P*P))
C     FOLDING
      J = 6002-I
      C(J) = C(I)
  775 W = W+2.E-6
      CALL LGRAPH(C(2501),1001,1,5)
C
C     CONVOLUTION
      DO 4 M = 1,4001
      K=M-2001
      SUM(M) = 0.0
      DO 5 I = 2001,4001
      D = E(I)*C(I-K)
    5 SUM(M) = SUM(M) + D
    4 SUM(M) = SUM(M)/2001.0
      CALL LGRAPH(SUM(1001),2001,1,10)
C     C IS FILTER SHAPE WIDENED BY FP PROFILE (7.5 A)
C     (OVERWRITTEN ON FP PROFILES)
      V = 2.77260/((7.5*50.0)**2)
      DO 68 K = 1001,3001,1
      Y = FLOAT(K-2001)
   68 C(K) = 100.0*EXP(-V*Y*Y)
C     DIVIDE BY FILTER AS IN REDUCTION
      DO 97 M = 1,3001
   97 SUM(M) = SUM(M)/C(M)
      CALL LGRAPH(SUM(1751),501,1,5)
  666 FORMAT (20X,13H   FINESSE = ,F4.0/20X,13HSEPARATION = ,F5.0
     1 ,8H MICRONS//7X,11HWHICH GIVES//11X,22HFREE SPECTRAL RANGE = ,
     2 F7.3,9HANGSTROMS/16X,17HRESOLVED WIDTH = ,F7.3,10H ANGSTROMS)
      WRITE (6,666)F,T,FREE,R
      GO TO 774
   25 STOP
      END
```

Appendix 5 Hicks, May and Reay NATURE paper 1972

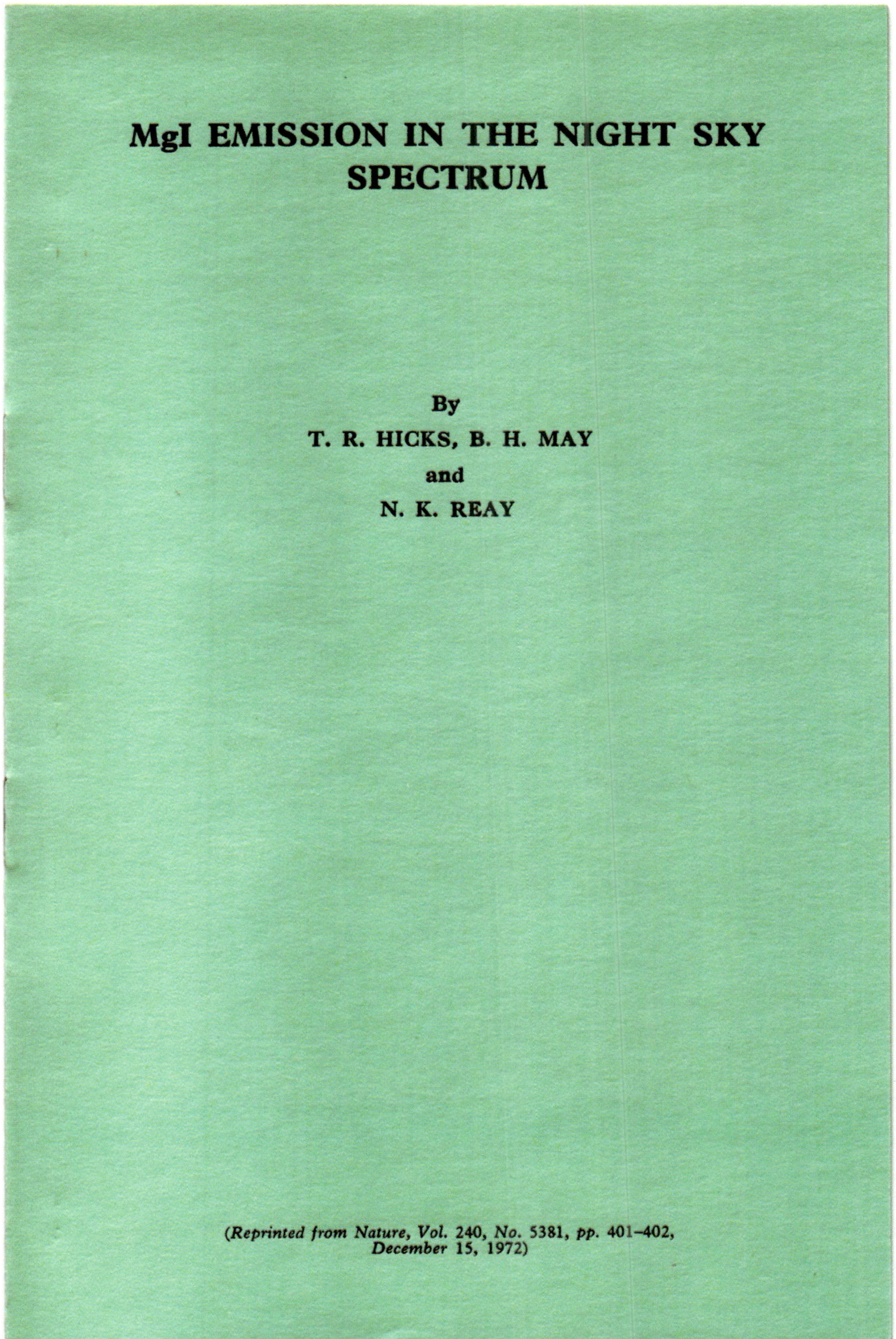

MgI EMISSION IN THE NIGHT SKY SPECTRUM

By

T. R. HICKS, B. H. MAY

and

N. K. REAY

(Reprinted from Nature, Vol. 240, No. 5381, pp. 401–402, December 15, 1972)

(Reprinted from Nature, Vol. 240, *No.* 5381, *pp.* 401–402,
December 15, 1972)

MgI Emission in the Night Sky Spectrum

DURING September, October 1971 and April 1972 we observed a narrow region of the night sky spectrum in the vicinity of the 5183.62 Å MgI wavelength. Measurements were made with a scanning Fabry-Perot interferometer[1,2] from the observatory at Izaña on Tenerife, Canary Islands, and consisted of spectral scans of up to 4 Å wide centred on the MgI wavelength. In all, some 350 spectra were obtained. Because the primary aim of

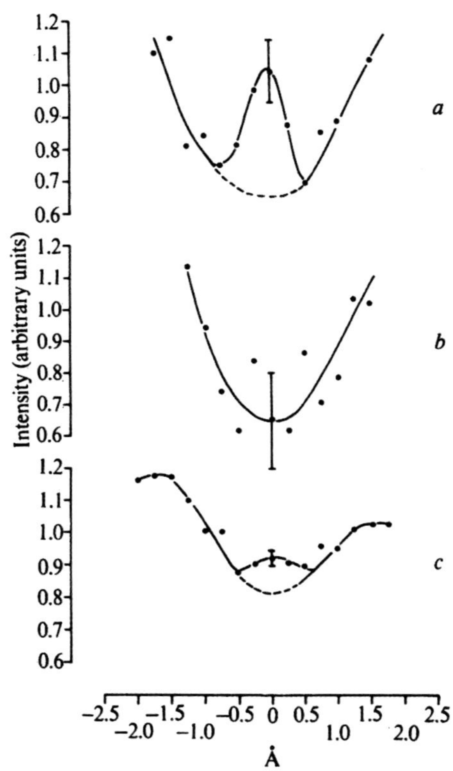

Fig. 1 Typical scans obtained in April 1972. *a* and *b*, Individual scans illustrating the quality of data (see text). *c*, Average of 7 scans recorded during the night of April 4/5. This clearly shows a significant emission.

the experiment was to study the zodiacal light, all spectra were recorded between evening and morning twilights, and most in the ecliptic plane. None was recorded during twilight.

The method of observation was to sample, for 48 s, each of up to 18 points across the spectral interval. Pulse counting electronics and a line printer recorded the signal level at each sample point. A second channel of pulse counting monitored the overall sky background over a wide waveband, thus allowing correction for fluctuations in sky transparency. The resolving power of the interferometer was 3,500, corresponding to an instrumental profile width of 1.5 Å.

Careful examination of the spectra revealed the presence of a small, variable, but undoubtedly real emission core at the centre of the zodiacal light Fraunhofer line. In general, because of the low signal levels, individual scans were not always sufficiently good to make measurements from, but scan averaging provided data of excellent quality. Fig. 1*a* shows a typical scan clearly showing emission, and 1*b* a scan in which there is no clear evidence for emission.

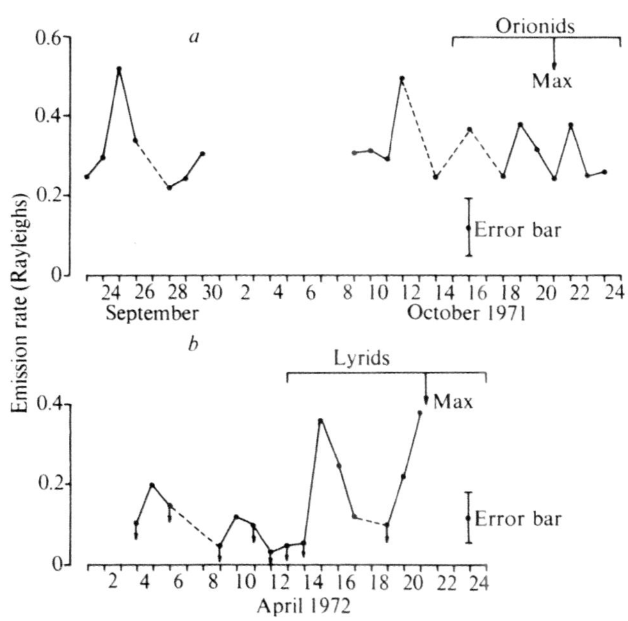

Fig. 2 Mean night-time emission rate as a function of time (see text). *a*, September, October 1971 data show little significant variation from night to night. The gap in data coincides with the period near full Moon. *b*, April 1972 data show emission rates lower than in *a*, with features coinciding in time with the Lyrid meteor shower. Vertical arrows on points denote upper limits.

The wavelength of the emission was determined by calibrating the interferometer with a magnesium hollow cathode lamp. The emission peak could be fixed to better than ± 0.2 Å from those individual scans which clearly showed emission. By combining the measurements from all such scans, the emission wavelength can be fixed at 5183.6 Å $\pm$ 0.1 Å. Reference to the MIT wavelength tables[3] shows that lines of cobalt, europium, zirconium and magnesium occur very near to 5183.6 Å. We identify the emission as being the 5183.618 Å line arising from the $3p^3P_2$–$4s^3S$ transition of MgI.

The well established enhancement of CaII emission at times of high meteor activity[4] led us to look for a correlation between MgI emission and meteor showers. The Orionid shower has a maximum in activity around October 20 and the Lyrid shower about April 21 (ref. 5).

Averaging of the data for a complete night of observation for all parts of the sky allowed mean emission levels to be obtained (Fig. 1*c*). Fig. 2*a* and *b* show emission rates, at the base of the atmosphere, derived from such averages for the two observing periods. Emission rates (in Rayleighs) are derived from a photometric calibration of the instrument; they are probably good to a factor of 4. Measurements at the "*V*" wavelength give extinctions of typically 0.15 mag. in the zenith for the Izaña site[6].

The September and October 1971 data show little significant variation in emission from night to night, having a mean level of about 0.3 Rayleigh.

The April 1972 data show a generally lower emission rate of about 0.1 Rayleigh, with some interesting structure beginning on the night of April 14/15, coinciding with the onset of the Lyrid meteor shower[5]. Two scans recorded near morning twilight on April 15 showed an unusually strong emission of about 0.7 Rayleigh. This declined over the next few nights until another enhancement occurred on April 20/21 corresponding to the time of maximum activity of the Lyrid shower. Unfortunately our data do not extend beyond April 21. It is worth noting that the "enhancements" in late April 1972 merely bring the emission rate up to the September, October 1971 level. In this respect it is not clear whether the emission undergoes an "enhancement" in late April, or is merely at a depressed level in early April. We believe the enhancement to be real, but the difference in ambient emission rates between September, October 1971 and April 1972 may be an artefact of the calibration technique, and modifications to the instrument fore-optics.

Enhancement of the emission at the time of the Lyrid meteor shower suggests that at least part of the atmospheric magnesium is of meteoric origin. No such enhancement occurred during the October 1971 Orionid shower, however.

Arising from the triple $3p^3P$–$4s^3S$ transition the emission cannot be due to fluorescence excitation, and indeed is observed well into the night when the 90–100 km region of the atmosphere, where most other metallic ions are observed[7,8], is completely in shadow.

Radiative recombination of Mg^+ with an electron, the mechanism suggested by Anderson and Barth[9] governing the daytime balance between Mg and Mg^+, is the most likely cause of the emission. It would be expected that other MgI lines, such as the other lines of the $3p^3P$–$4s^3S$ triplet 5172.68 Å and 5167.33 Å, are present with 5183.62 Å. The intercombination line at 4571.1 Å connecting ^3P to the ground ^1S state should also be present. We know of no observational searches for these emission lines.

This work was carried out in part with the aid of a University of London Central Research Fund grant. Two of us (T. R. H. and B. H. M.) gratefully acknowledge support in the form of SRC studentships.

We thank Professor Francisco Sanchez Martinez and Dr Carlos Sanchez Magro, of the University of La Laguna, for their invaluable assistance during our stay in Tenerife.

T. R. HICKS
B. H. MAY
N. K. REAY

Astronomy Group, Physics Department, Imperial College London

Received August 21, 1972.

1 Hindle, P., Reay, N., and Ring, J., *J. Sci. Instr.*, **44**, 646 (1967).
2 Hindle, P. H., and Reay, N. K., *J. Sci. Instr.*, **44**, 360 (1967).
3 *M.I.T. Wavelength Tables* (Wiley, New York, 1939).
4 Vallance Jones, A., *Nature*, **178**, 276 (1956).
5 Lovell, A. C. B., *Meteor Astronomy* (Oxford University Press, London, 1954).
6 McInnes, B., ROE Memorandum *Site Testing Project*, 1972.
7 Narcisi, R. S., *Space Research*, **8**, 360 (1968).
8 Narcisi, R. S., and Bailey, A. D., *J. Geophys. Res.*, **70**, 3687 (1965).
9 Anderson, J. G., and Barth, C. A., *J. Geophys. Res.*, **76**, 3723 (1971).

Printed in Great Britain by Flarepath Printers Ltd., St. Albans, Herts.

Appendix 6 Hicks, May and Reay MNRAS paper 1974

AN INVESTIGATION OF THE MOTION OF ZODIACAL DUST PARTICLES – I. RADIAL VELOCITY MEASUREMENTS ON FRAUNHOFER LINE PROFILES

T. R. Hicks, B. H. May and N. K. Reay

REPRINTED FROM

Mem. R. astr. Soc. (1974) **166,** 439–448

PUBLISHED BY

BLACKWELL SCIENTIFIC PUBLICATIONS

OXFORD LONDON EDINBURGH MELBOURNE

Mon. Not. R. astr. Soc. (1974) **166**, 439–448.

AN INVESTIGATION OF THE MOTION OF ZODIACAL DUST PARTICLES—I

Radial Velocity Measurements on Fraunhofer Line Profiles

T. R. Hicks, B. H. May and N. K. Reay

(Communicated by J. Ring)

(Received 1973 August 3)

SUMMARY

An experiment to record the spectrum of the Zodiacal Light in the neighbourhood of the Mg I absorption line (5183·6 Å) is described. Measurements were made of the Doppler shift imposed on the absorption line by the motion of the interplanetary dust particles. Observations were concentrated on the ecliptic plane, spectra being obtained at lower elongation angles from the Sun than previously achieved, and also over the entire range of high elongations including the Gegenschein.

The reduction methods applied to the data are described and compared, and the new results are presented.

1. INTRODUCTION

Several attempts have been made to measure the wavelength shifts, imposed by the motions of interplanetary dust particles on scattered solar Fraunhofer lines, in the Zodiacal Light (1)–(3).

The observations were made by examining the Zodiacal Light spectrum over a few Ångström units centred on a suitable Fraunhofer line, with high luminosity Fabry–Perot interferometers. Initially the Hβ 4861 Å line was used by Clarke *et al.* (1) and Reay & Ring (2), but a strong emission core detected at this wavelength (4), (5) caused James & Smeethe (3) to use the Mg I 5183·6 Å line. Because of the faintness of the Zodiacal Light, few accurate measurements have been made at elongations greater than 50°.

Models of the Zodiacal dust cloud have been constructed (6), (7) based on the available low elongation data, but the level of confidence placed in such models cannot be high until accurate data is obtained over the entire range of elongations.

For the last three years observations have been conducted by the Astronomy Group of Imperial College to obtain such accurate data, with the ultimate aim of constructing a model of the dust cloud. This follows a similar approach to that used by Reay & Ring (2) and later James & Smeethe (3). The measurements were made on the Mg I 5183·6 Å line from the ‘ Observatorio del Teide ’, Tenerife. Stringent precautions were taken to minimize errors in wavelength measurement and a large body of accurate data was collected, about 300 scans in all, during 1971 September–October and 1972 April.

2. THE INSTRUMENT

Light was directed into the laboratory using a 6-in. aperture two-mirror coelostat, with a third mirror to produce a horizontal beam 48 in. from the floor.

The light was then focused onto the entrance aperture of a gas pressure-scanned Fabry–Perot interferometer.

The 4-in. aperture Fabry–Perot interferometer (**8**) incorporates a fully automatic gas pressure stepping and cycling control system with which the wavelength points to be sampled are selected on a programmable switchboard. Nitrogen gas was used to scan the instrument allowing a maximum spectral range of 9 Å units, corresponding to 6 atm of pressure. The resolving power was 3400. Fig. 1 is a schematic diagram of the instrument.

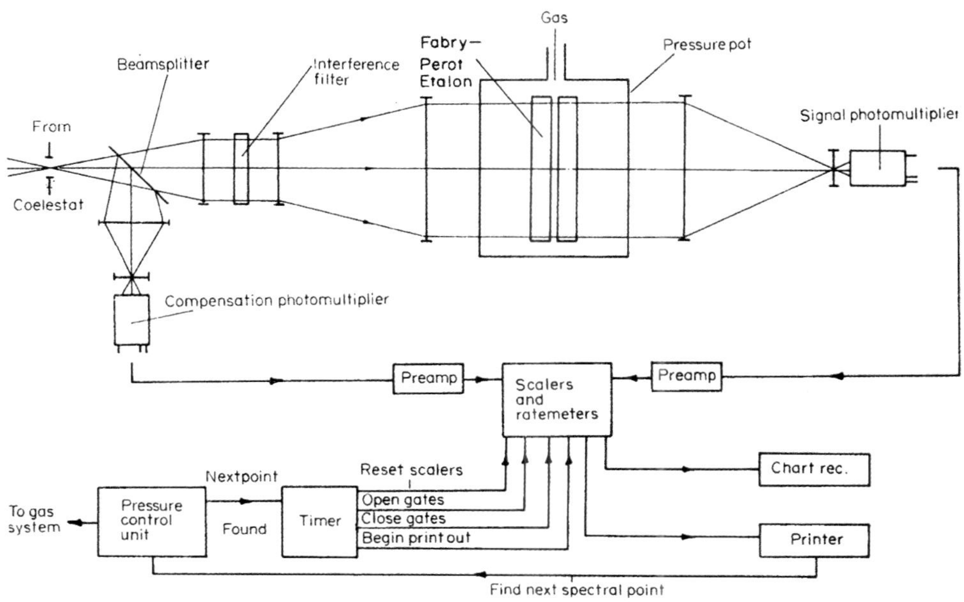

FIG. 1. *Schematic diagram of the instrument. The Pressure Control Unit automatically sets and stabilizes the gas pressure at the correct value for the next spectral point. The scalers then count for a preset time and the counts are recorded on both chart recorder and line printer. After print-out the controller is advanced by a pulse from the printer.*

Compensation for sky transparency fluctuations was provided for by diverting a few per cent of incident light through a 400 Å wide filter, centred on the Mg I wavelength, into a second channel of pulse counting electronics. The dark counting rate of the EMI 6256S signal channel photomultiplier tube was reduced to about 1 per second by cooling it to $-30\,^{\circ}$C using a Peltier-cooled housing. This compared with a typical sky counting rate of 5 per second.

Outer orders of the Fabry–Perot profile were suppressed using a 6 Å passband dielectric interference filter.

3. OBSERVATIONS AND DATA REDUCTION

Observations were concentrated to the ecliptic plane, a number of suitable points free from nebulosity and bright stars being previously tabulated in R.A. and declination.

The technique was to choose one of these tabulated points, set the coelostat to view this point, and sample between 10 and 18 spectral elements over a range of about 4 Å centred on the 5183·6 Å Mg I Fraunhofer line. The integration time for each sample element was 50 s. On each night the points available were systematically covered, using the periods close to twilight to look at low elongations. From night to night, of course, the tabulated points shifted in elongation from the Sun and so a fairly uniform sampling of the ecliptic was achieved, with a deliberate concentration on the Gegenschein. The area of the sky projected onto the entrance aperture was about one square degree, and since the duration of a scan was typically 12 min the spatially-resolved patch on the sky was an oval 3° long by 1° wide.

Calibration

The wavelength scale was established experimentally by scanning the 5183·6 Å line from a Magnesium lamp and the 5181·99 Å line from a Zinc lamp. The separation of these two lines in pressure allowed a wavelength scale to be constructed.

Data reduction

A typical ' raw ' data scan is shown in Fig. 2(a). For each signal point a mean value of dark current at the time is subtracted and the resulting figure, the true signal count, is divided by the compensation channel counts, giving a correction for overall light level changes during the scan. This number is then divided by the filter transmission value at the wavelength of this element. The (normalized) result is shown in Fig. 2(d). The fall-off at the edges of the ' raw ' data scan is seen to have been entirely due to the interference filter, and the poissonian error bars are consequently larger for these end points.

A number of scans were recorded which showed an emission feature at the 5183·6 Å wavelength. This we interpreted as being atmospheric in origin and possibly associated with meteor showers. These data have been the subject of a previous communication (**9**) and are not included in the present discussion.

In the case of scans recorded near dawn it was possible to compensate for daylight contamination. The onset of twilight within 10° of the horizon (where much of our work at very low elongations was done) did not necessarily coincide with the beginning of Astronomical Twilight, but varied by as much as half an hour depending on the local atmosphere towards the east. A scan recorded at dawn is shown in Fig. 3(a). The beginning of daylight contamination, as opposed to the very slight slope due to the rising of the Zodiacal Light cone, is clearly visible in the compensation channel and it is reasonable to suppose that the shaded portion of Fig. 3(b) is mainly daylight. Solar scans were obtained during the day and so this spectrum is well known and virtually noiseless owing to high daytime fluxes.

If at any spectral point the fraction of daylight present in the scan is F, the number of counts to be subtracted from the appropriate scan point to eliminate daylight is $F.D(\lambda).C$ (Fig. 3(c)) where $D(\lambda)$ is the daylight spectrum scaled to unity and C is the assumed true continuum level in the spectrum under study. The corrected spectrum is shown in Fig. 3(d). The small magnitude of the corrections in the central region indicates that the method could be used to good effect even further into twilight and at smaller elongations.

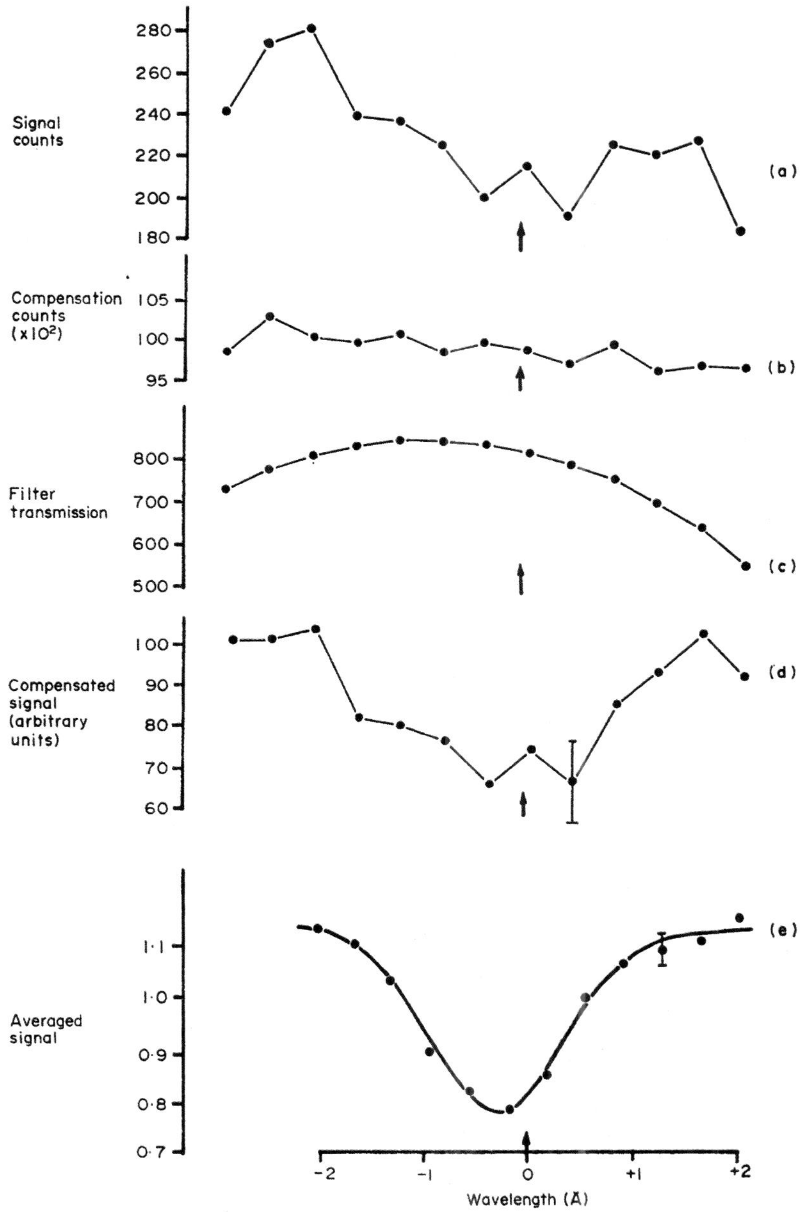

FIG. 2. *A typical scan showing: (a) ' Raw ' data from the signal channel. (b) Compensation channel data. (c) Filter profile. (d) The result of subtracting the ' dark ' count and dividing (a) by both (b) and (c) to compensate for filter transmission and atmospheric transparency fluctuations. (e) A typical averaged profile, derived in this case from 5 scans.*

The arrow indicates the nominal λ 5183·6 Å *position determined from calibration scans.*

The apparent central peak in the raw data is smaller than the error bar and therefore not significant.

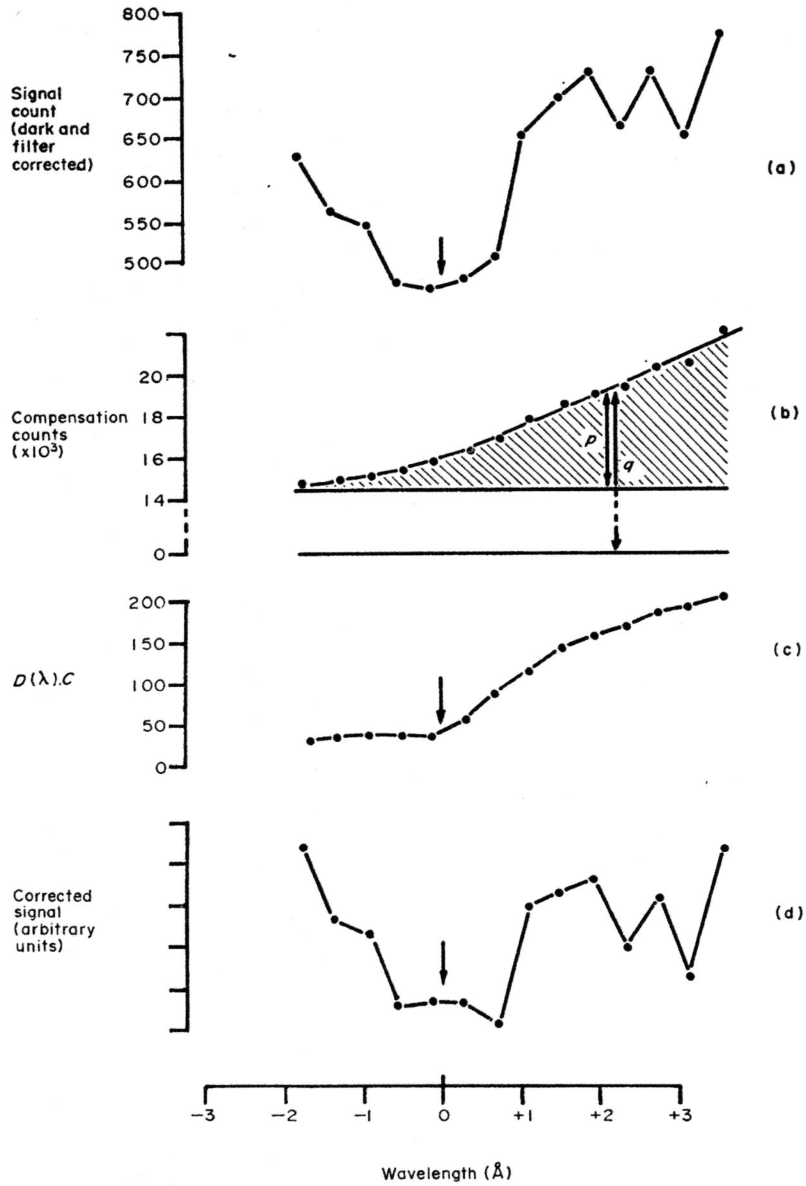

FIG. 3. *The method of correcting for daylight contamination: (a) Scan showing effects of twilight. (b) Compensation channel record; the shaded area represents the daylight contamination. The fraction of the count F contributed by daylight is p/q. (c) Computed contribution of daylight to signal. (d) Corrected Zodiacal Light scan.*

The data were processed as a function of elongation, that from the two periods being treated separately in an effort to untangle temporal effects or directional effects on the Celestial Sphere (see review by Divari (**10**)).

The ecliptic plane was divided into 10° intervals (5° for low elongations) and an average spectrum was calculated for each group of scans in an interval. Such an average is shown in Fig. 2(e). Most averages indicated a well-behaved absorption feature, and various curve fittings were made to obtain the best estimate of the absorption line minimum.

Initially a fourth order polynomial fit was made to the data points and provided a good fit in most cases. The analytical form of the polynomial yielded immediately two crude guides to the position of the line; the minimum point, and the bisector of the line joining the two points of inflexion. We could also find an ordinate which bisected the area of the function below a nominal continuum level.

A necessary constraint on the method used to estimate the wavelength of the absorption feature is that it uses the information for *all* the points in the optimum way, bearing in mind their various probable errors, and that the measure be identifiable with a quantity calculable from models of the dust cloud. In practice this quantity is the ' mean shift ':

$$\langle\Delta\lambda\rangle = \frac{\int \Delta\lambda . dB}{\int dB}$$

where $\Delta\lambda$ is the shift due to an element of brightness dB and the integral is along a line of sight (**6**), (**7**). If the Zodiacal Light spectrum is symmetrical (a good approximation at 1·5 Å resolution) it is easy to show that $\langle\Delta\lambda\rangle$ is identical to the shift of the bisector of the line's area. $\langle\Delta\lambda\rangle$ is also identical to the shift of the bisector of the inflexion points, and to the minimum point.

Noise in the practical spectrum reduces the reliability of all three methods. The minimum point method is too sensitive to the few central points, and a small emission, as was sometimes present, is enough to upset the measure. The inflection method is too sensitive to the noisy outer parts of the spectrum. The method of areas would rely on a good estimate of the continuum level, which again depends strongly on the outer points. A small error in the placing of the filter can produce ana symmetry in the continuum levels.

The method finally adopted was to fit a Gaussian curve

$$I(\lambda) = V_1 - V_2 \exp\left[-V_3(\lambda-\Delta\lambda)^2\right] \tag{1}$$

to the experimental points.

This is justified experimentally in that the fit obtained is not highly sensitive to the presence of a small central emission core, or to edge effects; also in that the eventual residuals are small.

It is justified theoretically by considering the line shape expected from model dust clouds by making ' line of sight ' integrations. The calculated absorption feature, when convolved with the instrumental function, is symmetric and compares well with a Gaussian (Fig. 4).

The parameter $\Delta\lambda$ in equation (1) is the shift of the line from its nominal position and the line half width W is

$$2\left[\frac{\log_e 2}{V_3}\right]^{1/2}.$$

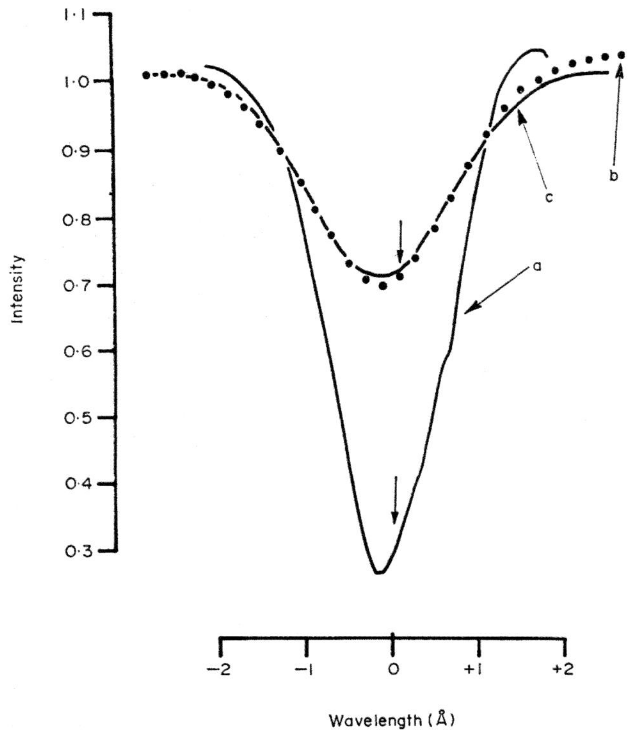

FIG. 4. *Comparison of a Gaussian with theoretical profiles. (a) A 30° elongation profile derived for a model of the Zodiacal dust cloud which is co-rotating with the Earth in circular orbits. The dust number density is assumed to fall away from the Sun as r^{-1}. (b) This profile convolved with the instrument function of our Fabry–Perot interferometer. (c) A comparison Gaussian curve.*

The Doppler shifts obtained from the Gaussian fits to the data are plotted in Fig. 5 as a function of elongation.

4. DISCUSSION

The data for the two observing periods appear generally consistent in that both sets show negative and positive radial velocity maxima in the morning and evening respectively. Detailed differences occur however which are best illustrated by reference to Fig. 5, which shows third order polynomial fits to both sets of data.

A distinct morning–evening asymmetry occurs in the 1971 September–October data, with an apparent recession of the Gegenschein from the Earth at a speed $\sim 12\ \mathrm{km\,s^{-1}}$ (0·2 Å). The evening radial velocity maximum at elongation 110° is significantly higher than the morning maximum at elongation 40°.

The 1972 April data is symmetric about the antisolar point to within experimental error. In general this data is in agreement with that of Reay & Ring (2) and James & Smeethe (3) at *low* elongations. At *high* elongations our new data disagrees substantially with the previous data, plotted on Fig. 5 for comparison.

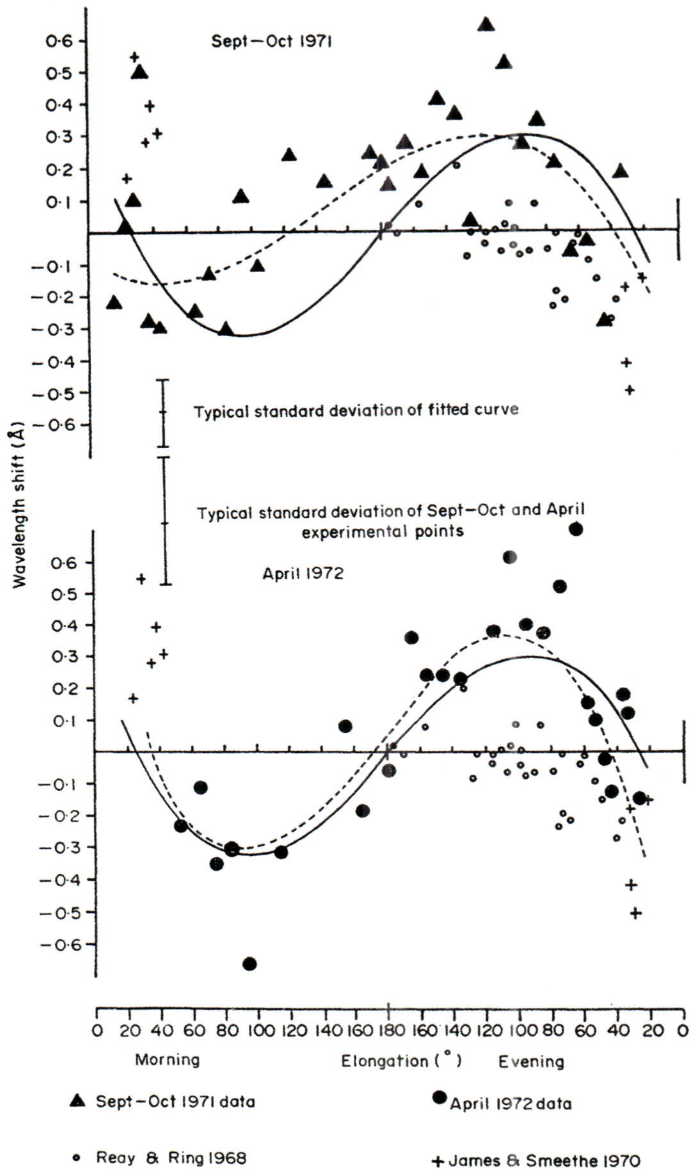

FIG. 5. *Doppler shift as a function of elongation in both morning and evening for the two periods of observation: (a) 1971 September–October; and (b) 1972 April. Weighted third order polynomial least squares fits (------), to each set of data, are shown compared with a theoretical curve (——) derived from the prograde orbital dust model of Reay (7) (see text). Previous data of Reay & Ring (2) and James & Smeethe (3) are shown for comparison.*

Reay (7) attempted to explain the early data in terms of dust in circular gravitational orbits about the Sun and co-rotating with the Earth and planets. On Fig. 5 we have plotted the expected radial velocity behaviour for such a model in which the dust density falls as r^{-2} from the Sun and the distribution of particle sizes/cm^3 is given by:

$$n(a)\,da = ka^{-p}\,da,$$

where

$$\begin{aligned} a &= \text{particle radius,} \\ n(a) &= \text{number/cm}^3 \text{ with radius ` } a \text{ ',} \\ k &= \text{constant.} \end{aligned}$$

The curve was computed assuming:

$$\begin{aligned} \text{minimum particle radius } a_1 &= 0{\cdot}2\ \mu\text{m}, \\ \text{maximum particle radius } a_2 &= 100\ \mu\text{m}, \\ \text{distribution parameter } p &= 5, \\ \text{particle albedo } \gamma &= 0{\cdot}2, \\ \text{particle density } \rho &= 3\ \text{g/cm}^{-3}. \end{aligned}$$

Radiation pressure and diffraction have been taken into account in the calculation.

Comparison with the polynomial fits show that this model agrees well with the 1972 April data. Deviations of the 1971 September–October data from the model are significant over the entire range of elongations and suggest the model to be totally inadequate in this case.

The apparently large scatter in the morning elongation interval 30°–40°, for 1971 September–October, makes the validity of the polynomial fit in this region uncertain. Whether or not the scatter is a real time variation (the scans averaged to provide these points were obtained over a period of several weeks) will be the subject of further investigation.

Inclusion in the calculation of elliptic orbits may possibly improve the correspondence between data and model, particularly if an uneven distribution of orbital perihelia, with ecliptic longitude, can be assumed. The attraction of Jupiter has been suggested as a possible cause of such a distribution (6).

An alternative approach, which we are investigating, is that the shape of the radial velocity *vs.* elongation curve in 1971 September–October was influenced at large elongation by interstellar dust streaming through the solar system.

In summary we would suggest that a high level of confidence can be placed in the data presented above. Care has been taken to eliminate sources of error in wavelength determination. The data provide a comprehensive coverage of the ecliptic plane for both morning and evening elongations, and should prove useful in the construction of models of the Zodiacal dust cloud.

ACKNOWLEDGMENTS

The work was supported in part by the ' University of London Central Research Fund '.

We wish to thank Professor J. Ring of Imperial College for his constant support for this project, also Professor Francisco Sanchez Martinez and Dr Carlos Sanchez

Magro of the University of La Laguna for their assistance and cooperation during our stay in Tenerife.

Two of us (TRH and BHM) were in grateful receipt of SRC studentships at the time this research was carried out.

Astronomy Group, Physics Department, Imperial College of Science and Technology, London SW7 1NA

REFERENCES

(1) Clarke, D., Daehler, M., Mack, J. E., Ring, J. & Stoner, J. O., 1967. *Modern Astrophysics*, ed. M. Hack, Gauthier-Villars, Paris.
(2) Reay, N. K. & Ring, J., 1968. *Nature*, **219**, 710.
(3) James, J. F. & Smeethe, M. J., 1970. *Nature*, **227**, 589.
(4) Daehler, M., Mack, J. E., Stoner, J. O., Clarke, D. & Ring, J., 1968. *Planet. Space Sci.*, **16**, 795.
(5) Reay, N. K. & Ring, J., 1969. *Planet. Space Sci.*, **17**, 561.
(6) Bandemann, L. W. & Wolstencroft, R. D., 1969. *Nature*, **221**, 251.
(7) Reay, N. K., 1969. *Nature*, **224**, 54.
(8) Hindle, P. H. & Reay, N. K., 1967. *J. Sci. Instrum.*, **44**, 360.
(9) Hicks, T. R., May, B. & Reay, N. K., 1972. *Nature*, **240**, 401.
(10) Divari, N. B., 1965. *Sov. Phy., Uspeki*, **7**, 681.

End of Hicks, May, Reay paper, MNRAS 1974

Appendix 7 Derivation of the Airy Summation for a Fabry-Perot Interferometer

The diagram shows two rays emerging from the Fabry-Perot, one of which, T_0, passes through the etalon without reflections, and the other, T_1, is reflected twice by the coatings, before emerging. The refractive index of the medium between the plates is n, and that of the medium outside the gap is n_0. In our case the medium outside the gap is the material of the plates themselves, quartz, and the emerging rays have to pass through another refraction before they finally leave the etalon, but the pattern of the resulting transmission is unaffected by this, so it is ignored in this analysis.

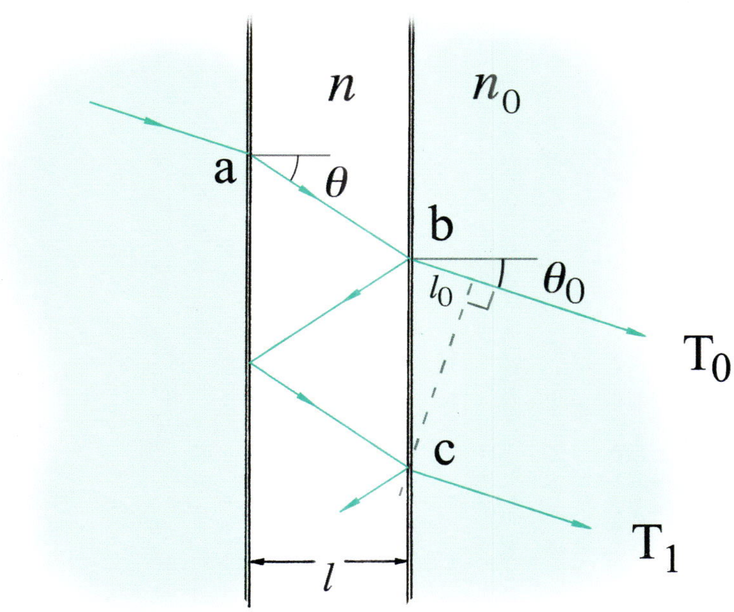

At each reflection, the amplitude is reduced by $\sqrt{R}$, and the phase is shifted by π, while at each transmission through an interface the amplitude is reduced by $\sqrt{T}$. Assuming no absorption, we have, by the law of conservation of energy, $T + R = 1$.

Using complex form for a light ray, and defining the amplitude of the wave at point **(a)** as unity, the amplitude at point **(b)** is

$$T_0 = T\exp(ikl/\cos\theta)$$

where k is the wave number inside the etalon, given by $k = 2\pi\, n/\lambda$, and λ is the vacuum wavelength. At point **(c)** the amplitude is

$$T_1 = TR\,\exp[(2\pi i + 3ikl)/\cos\theta]$$

The total amplitude of the two beams will be the sum of their amplitudes, measured at a wavefront defined by a line perpendicular to the direction of the beam. We therefore need to add, to the amplitude at point **(b)**, a modified amplitude, equal in magnitude to T_1, but retarded in phase by an amount $k_0 l_0$, where k_0 is the wave number outside of the etalon, given by $k_0 = 2\pi n_0/\lambda$. Thus

$$T_1 = RT\exp[(2\pi i + 3ikl)/(\cos\theta - ik_0 l_o)]$$

where l_0 is given by $l_0 = 2l\tan\theta\,\tan\theta_0$

Neglecting the 2π phase change due to the two reflections, we have for the phase difference between the two beams

$$\delta = 2kl/\cos\theta - k_0 l_0$$

From the law of refraction

$$n\sin\theta = n_o\sin\theta_0$$

So the phase difference may be written

$$\Delta = 2kl\cos\theta$$

To within a constant multiplicative phase factor, the amplitude of the mth emergent beam can be written as

$$T_m = TR^m\exp(im\theta)$$

The total transmitted beam is the sum of all individual beams

$$A_T = \sum_{m=0}^{\infty} T_m = T \sum_{m=0}^{\infty} R^m \exp(im\Delta)$$

This is the sum of a geometric series, and so can be written as

$$A_T = T/[1 - R\exp(i\Delta)]$$

The intensity of the beam is now $A_T A_T^*$ and, since the incident beam was assumed to have an intensity of unity, this will also give the transmission function:

$$\tau = \frac{T^2}{1 + R^2 - 2R\cos\Delta}$$

This completes the derivation of equation 2.1, page 55.

References

Aumann, H. H.; Beichman, C. A.; Gillett, F. C.; de Jong, T.; Houck, J. R.; Low, F. J.; Neugebauer, G.; Walker, R. G.; Wesselius, P. R. Discovery of a shell around Alpha Lyrae. Astrophysical Journal, Part 2 - Letters to the Editor (ISSN 0004-637X), Vol. 278, March 1, 1984, pp. L23-L27 (1984).

Backman, Dana E.; Paresce, Francesco. Main-sequence stars with circumstellar solid material - The Vega Phenomenon. In: Protostars and Planets III (A93-42937 17-90), pp. 1253-1304 pub: University of Arizona Press (1993).

Baggaley, W. J. Advanced Meteor Orbit Radar (AMOR) observations of interstellar meteoroids. Journal of Geophysical Research, Vol. 105, Issue A5, pp. 10353-10362 (2000).

Baguhl, M.; Grün, E.; Landgraf, M. In Situ Measurements of Interstellar Dust with the Ulysses and Galileo Spaceprobes. Space Science Reviews, Vol. 78, Issue 1/2, pp. 165-172 (1996).

Bandermann, L. W.; Wolstencroft, R. D. Interpretation of Doppler Shifts in the Zodiacal Light Spectrum. Nature, Vol. 221, Issue 5177, pp. 251-253 (1969).

Beavers, W. I.; Eitter, J. J.; Carr, P. H.; Cook, B. C. Radial velocity measurements in the F-Corona. Astrophysical Journal, Part 1, Vol. 238, pp. 349-356 (1980).

Behr, A.; Siedentopf, H. Untersuchungen über Zodiakallicht und Gegenschein nach lichtelektrischen Messungen auf dem Jungfraujoch. Zeitschrift für Astrophysik, Vol. 32, p.19 (1953).

Beichman, C. A. The IRAS view of the Galaxy and the Solar System. Annual review of astronomy and astrophysics, Vol. 25 (A88-13240 03-90). Palo Alto, CA, Annual Reviews, Inc., pp. 521-563 (1987).

Belton, Michael J. S. Dynamics of Interplanetary Dust. Science, Vol. 151, Issue 3706, pp. 35-44 (1966).

Bernstein, R. A.; Freedman, W. L.; Madore, B. F. The First Detections of the Extragalactic Background Light at 3000, 5500, and 8000 Å. II. Measurement of Foreground Zodiacal Light. Astrophysical

Journal, Vol. 571, pp. 85-106 (2002).

Blackwell, D. E.; Ingham, M. F. Observations of the zodiacal light from a very high altitude station: II. Electron densities in interplanetary space. Monthly Notices of the Royal Astronomical Society, Vol. 122, p. 129 (1961).

Bock, J.; Battle, J.; Cooray, A.; Kawada, M.; Keating, B.; Lange, A.; Lee, D-H.; Matsumoto, T.; Matsuura, S.; Pak, S.; Renbarger, T.; Sullivan, I.; Tsumura, K.; Wada, T.; Watabe, T. The Cosmic Infrared Background Experiment (CIBER). New Astronomy Reviews, Vol. 50, Issue 1-3, pp. 215-220 (2006).

Brorsen, T. Über eine neue erscheinung am Zodiacallichtes, geographie und Witterungskunde, Vol. 8, pp. 156-60 (1854).

Cassini, G. D. Découverte de la lumiere celèste qui paroist dans le Zodiac. Memoires de l'Académie Royale des Sciences depuis 1666 jusqu'a 1699, Tome VIII, Paris 1699. (Compagnies des Libraires, 1730), pp. 119-209 (1699).

Childrey, J. Britannica Baconica, or the Natural Rarities of England, Scotland, and Whales *(sic)*, London (1661).

Clarke, D.; Matthews, S. A.; Mundell, C. G.; Weir, A. S. On the line profiles in the spectra of the Zodiacal Light. Astronomy and Astrophysics, Vol. 308, pp. 273-278 (1996).

Crovisier, J.; Leech, K.; Bockelee-Morvan, D.; Brooke, T. Y.; Hanner, M. S.; Altieri, B.; Keller, H. U.; Lellouch, E. The Spectrum of Comet Hale-Bopp (C/1995 01) observed with the Infrared Space Observatory at 2.9 AU from the Sun. Science, Vol. 275, pp. 1904-1907 (1997).

Daehler, M.; Mack, J. E.; Stoner, J., Jr.; Clarke, D.; Ring, J. Measurements of the Hβ profile in the Zodiacal Light spectrum. Planetary and Space Science, Vol. 16, p. 795 (1968).

Dermott, S. F.; Nicholson, P. D.; Wolven, B. In: Asteroids, Comets, Meteors II, eds: Lagervist, C.-I.; Rickman, H. Uppsala (1985).

Dermott, S. F.; Nicholson, P. D.; Kim, Y.; Wolven, B. In: Comets to Cosmology, ed: Lawrence, A. Springer-Verlag, Berlin p. 3 (1987).

Dermott, S. F.; Nicholson, P. D. IRAS dust bands and the origin of the zodiacal cloud. Highlights of Astronomy, Vol. 8, pp. 259-266 (1989).

Dermott, S. F.; Durda, D. D.; Gustafson, B. A. S.; Jayaraman, S.; Xu, Y.-L.; Gomes, R. S.; Nicholson, P. D. The origin and evolution of the zodiacal dust cloud. In: Lunar and Planetary Inst., Asteroids, Comets, Meteors 1991, pp. 153-156 ISBN: 0-942862-07-04 (1992).

Dermott, S. F.; Jayaraman, S.; Xu, Y.-L.; Liou, J.-C. IRAS observations show that the Earth is embedded in a solar ring of asteroidal dust particles in resonant lock with the planet. In: Lunar and Planetary Inst., Workshop on Particle Capture, Recovery and Velocity/Trajectory Measurement Technologies, pp. 35-36 (N95-19172 05-90) (1994).

Dermott, S. F.; Jayaraman, S.; Xu, Y.-L.; Grogan, K.; Gustafson, B. A. S. In: Unveiling the Cosmic Infrared Background, ed. E. Dweck. (New York: AIP), p. 25 (1996).

Divari, N. B. Astronomicheskii Zhurnal, Vol. 32 (1) p. 79 (1955).

Divari, N. B.; Asaad, A. S. Photoelectric Observations of the Zodiacal Light in Egypt. Soviet Astronomy, Vol. 3, p. 832 (1959).

Divari, N. B. Contribution of the Dust Cloud Near the Earth to the Brightness of the Zodiacal Light and of the F-Corona. Soviet Astronomy, Vol. 9, p. 493 (1965).

Divari, N. B. The Cosmic Dust Cloud around the Earth. Soviet Astronomy, Vol. 10, p. 1017 (1967).

Dobrovol'Skii, O. V.; Egibekov, P.; Zausaev, A. F. Nongravitational effects on the evolution of dust particles moving in elliptic orbits about the sun. Astronomicheskii Zhurnal, Vol. 50, p. 832 (1973).

Doppler, Christian; Studnica, Frantisek Josef. Über das farbige licht der doppelsterne und einiger anderer gestirne des himmels. Prag, K. Bohm gesellschaft der wissenschaften. Bibliographic code 1903 QB901.D69 (1903).

Draine, B. T.; Lee, H. M. Optical properties of interstellar graphite and silicate grains. Astrophysical Journal, Part 1 (ISSN 0004-637X), Vol. 285, pp. 89-108 (1984).

East, I. R.; Reay, N. K. Astronomy and Astrophysics Vol. 139, No. 2, pp. 512-516 (1984).

Elvey, C. T. Photo-electric Observations of the Color of the Zodiacal Light. Astrophysical Journal, Vol. 80, p. 61 (1934).

Elvey, C. T. and Roach, F. E. Astrophysical Journal, Vol. 85, p. 213 (1937).

Fechtig, H.; Leinert, C.; Berg, O. E. "Interplanetary Dust" (Springer, 2004).

Fernandez, J. A. On the observed excess of retrograde orbits among long-period comets. Monthly Notices of the Royal Astronomical Society, Vol. 197, pp. 265-273 (1981).

Fesenkov, V. G. Zodiacal Light as the Product of Disintegration of Asteroids. Soviet Astronomy, Vol. 2, p. 303 (1958).

Fixsen, D. J.; Dwek, E. The Zodiacal emission spectrum as determined by COBE and its implications. Astrophysical Journal, Vol. 578, pp.1009-1014 (2002).

Fried, J. W. Doppler Shifts in the Zodiacal Light Spectrum. Astronomy and Astrophysics, Vol. 68, pp. 229-264 (1978).

Gardner, James A.; Viereck, Rodney A.; Murad, Edmond; Knecht, David J.; Pike, C. P.; Broadfoot, A. Lyle; Anderson, Emmet R. Atmospheric Composition and Structure: Airglow and aurora, Ionosphere: Active experiments, Ionosphere: Equatorial ionosphere, Ionosphere: Ionospheric dynamics. Geophysical Research Letters, Volume 22, Issue 16, pp. 2119-2122 (1995).

Gemini Near Infrared Spectrograph website:
http://www.gemini.edu/sciops/instruments/nirs/nirsObsStrategies.html (2007).

Giese, R.H. Light scattering of small particles and models of interplanetary matter derived from the Zodiacal Light. Space Science Review, Vol. 1, pp. 589-611 (1963).

Gindilis, L. M. The Gegenschein as an Effect Produced by the Scattering of Light from Particles of Interplanetary Dust. Soviet Astronomy, Vol. 6, p. 540 (1963).

Gold, T. Resonant orbits of grains and the formation of satellites. (International Astronomical Union and COSPAR, Colloquium on Planetary Satellites, 28th, Cornell University, Ithaca, N.Y., Aug. 18-21, 1974). Icarus, Vol. 25, pp. 489-491 (July 1975).

Gorkavyi, N.; Ozernoy, L.; Mather, J.; Taidakova, T. NGST and the Zodiacal Light in the Solar System. Next Generation Space Telescope Science and Technology, ASP Conference Series, Vol. 207. Proceedings of a conference held at Hyannis, MA, 13-16 September, 1999. Edited by Eric Smith and Knox Long. ISBN: 1-58381-036-6 p. 462 (2000).

Griffin, R. F. A Photoelectric Radial Velocity Spectrometer. Astrophysical Journal, Vol. 148, p. 465 (1967).

Grogan, K; Dermott, S. F. and Gustafson, B. A. S. Astrophysical Journal, Vol. 472, p. 812 (1996).

Grogan, K.; Dermott, S. F.; Durda, D. D. The Size-Frequency Distribution of the Zodiacal Cloud: Evidence from the Solar System Dust Bands. Icarus, Vol. 152, Issue 2, pp. 251-267 (2001).

Grün, E.; Zook, H. A.; Baguhl, M.; Balogh, A.; Bame, S. J.; Fechtig, H.; Forsyth, R.; Hanner, M. S.; Horanyi, M.; Kissel, J.; Lindblad, B.-A.; Linkert, D.; Linkert, G.; Mann, I.; McDonnell, J. A. M.; Morfill, G. E.; Phillips, J. L.; Polanskey, C.; Schwehm, G.; Siddique, N.; Staubach, P.; Svestka, J.; Taylor, A. Discovery of Jovian dust streams and interstellar grains by the ULYSSES spacecraft. Nature, Vol. 362, No. 6419, pp. 428-430 (1993).

Grün, E.; Staubach, P.; Baguhl, M.; Hamilton, D. P.; Zook, H. A.; Dermott, S.; Gustafson, B. A.; Fechtig, H.; Kissel, J.; Linkert, D.; Linkert, G.; Srama, R.; Hanner, M. S.; Polanskey, C.; Horanyi, M.; Lindblad, B. A.; Mann, I.; McDonnell, J. A. M.; Morfill, G. E.; Schwehm, G. South-North and Radial Traverses through the Interplanetary Dust Cloud. Icarus, Vol. 129, Issue 2, pp. 270-288 (1997).

Grün, E.; Gustafson, B. A. S.; Dermott, S.; Fechtig, H. (eds) "Interplanetary Dust", pub: Springer (ISSN 0941-7834), (2001).

Hanner, M. S.; Weinberg, J. L. Gegenschein Observations from Pioneer 10. Sky and Telescope, Vol. 45, p. 217 (1967).

Hauser, M. G.; Gillett, F. C.; Low, F. J.; Gautier, T. N.; Beichman, C. A.; Aumann, H. H.; Neugebauer, G.; Baud, B.; Boggess, N.; Emerson, J. P. IRAS observations of the diffuse infrared background. Astrophysical Journal, Part 2 - Letters to the Editor (ISSN 0004-637X), Vol. 278, pp. L15-L18 (March 1984).

Hicks, T. R.; May, B. H.; Reay, N. K. Nature, Vol. 240, pp. 401-402 (1972).

Hicks, T. R.; May, B. H.; Reay, N. K. Monthly Notices of the Royal Astronomical Society, Vol. 166, pp. 439-448 (1974).

Hindle, P. H.; Ibbett, R. N. A digitization system for a scanning spectrometer. J. Sci. Instrum. Vol. 43, pp. 209-214 (1966).

Hindle, P. H.; Reay, N. K.; Ring, J. Hβ radiation in the spectrum of the night sky. Planetary and Space Science, Vol. 16, p. 803 (1968).

Hoffmeister, C. Das Spektrum des Nachthimmels und des Zodiakallichts nach Beobachtungen auf der Astronomischen Station bei Windhuk, Südwestafrika. Mit 7 Abbildungen. Zeitschrift für Astrophysik, Vol. 19, p. 116 (1940).

Hong, S. S. Henyey-Greenstein representation of the mean volume scattering phase function for Zodiacal Dust. Astronomy & Astrophysics (ISSN 004-6361), Vol. 146, No. 1, pp. 67-74 (1985).

Hubble Website: http://hubblesite.org/newscenter/newsdesk/archive/releases/2006/25/full/ (2007).

Huggins, W.; Miller, W. A. Note on the Lines in the Spectra of Some of the Fixed Stars. Proceedings of the Royal Society of London, Vol. 12, pp. 444-445 (1862).

Hulbert, E. O. Popular Astronomy, Vol. 38, p. 215 (1930).

Huruhata, Masaaki. Photoelectric observations of the photometric axis of zodiacal light. Planetary and Space Science, Vol. 13, p. 237 (1965).

Ingham, M. F. The nature and distribution of the interplanetary dust. Monthly Notices of the Royal Astronomical Society, Vol. 122, p. 157 (1961).

Ingham, M. F. The profile of an absorption line in the spectrum of the zodiacal light. Monthly Notices of the Royal Astronomical Society, Vol. 126, p. 377 (1963).

Ipatov, S. I.; Mather, J. C. Migration of small particles and dust to near-Earth space. Advances in Space Research, Vol. 37, pp. 126-137 (2005).

Ipatov, S. I.; Kutryev, A. S.; Madsen, G. J.; Reynolds, R. J. Dynamical Zodiacal Cloud Models constrained by High Resolution spectroscopy of the Zodiacal Light. Astro/ph 0608141 (August 2006).

Ishiguro, M.; Nakamura, R.; Fujii, Y.; Morishige, K.; Yano, H.; Yasuda, H.; Yokogawa, S.; Mukai, T. First Detection of Visible Zodiacal Dust Bands from Ground-based Observations. The Astrophysical Journal, Vol. 511, Issue 1, pp. 432-435 (1999).

Jacquinot, P. The Luminosity of Spectrometers with Prisms, Gratings or Fabry-Perot Etalons. J. Opt. Soc. Am., Vol. 44, p. 761 (1954).

James, J. F. Theoretical Fraunhofer line profiles in the spectrum of the zodiacal light. Monthly Notices of the Royal Astronomical Society, Vol. 142, p. 45 (1969).

James, J. F.; Smeethe, M. J. Motion of the Interplanetary Dust Cloud. Nature, Vol. 227, Issue 5258, pp. 588-589 (1970).

Jeffries, A. J. An investigation of the lunar ejecta theory of the cause of the Gegenschein. Planetary and Space Science, Vol. 19, p. 841 (1971).

Jessberger, E. K.; Stephan, T.; Rost, D.; Arndt, P.; Maetz, M.; Stadermann, F. J.; Brownlee, D. E.; Bradley, J. P.; Kurat, G. Properties of Interplanetary Dust: Information from Collected Samples. In Grün, E., Gustafson, B. A. S., Dermott, S. F., Fechtig, H. (eds.) Interplanetary Dust, pp. 253–294, Springer-Verlag (2001).

Jones, G. United States Japan Expedition, Vol. III, pub: Washington (1856).

Jones, M. H.; Rowan-Robinson, M. A Physical Model for the IRAS Zodiacal Dust Bands. Monthly Notices of the Royal Astronomical Society, Vol. 264, pp. 237-247 (1993).

Jones, O. C. Tables of Spectral Power Distribution. National Physical Laboratory, Division of Quantum Metrology Qu. 14 (November 1970).

Kelsall, T.; Weiland, J. L.; Franz, B. A.; Reach, W. T.; Arendt, R. G.; Dwek, E.; Freudenreich, H. T.; Hauser, M. G.; Moseley, S. H.; Odegard, N. P.; Silverberg, R. F.; Wright, E. L. The COBE Diffuse

Infrared Background Experiment Search for the Cosmic Infrared Background. II. Model of the Interplanetary Dust Cloud. The Astrophysical Journal, Vol. 508, Issue 1, pp. 44-73 (1998).

Kortenkamp, S. J.; Dermott, S. F. Accretion of Cometary Interplanetary Dust Particles. 29th Annual Lunar and Planetary Science Conference, Houston, TX, abstract No. 1618 (March 16-20, 1998).

Kurucz, R. L.; Furenlid, I.; Brault, J.; Testerman, L. Solar Flux Atlas from 296 to 1300 nm, NOAO Atlas No. 1 (1984).

Landgraf, M.; Baggaley, W. J.; Grün, E.; Krüger, H.; Linkert, G. Aspects of the mass distribution of interstellar dust grains in the solar system from in situ measurements. Journal of Geophysical Research, Vol. 105, Issue A5, pp. 10343-10352 (2000).

Leinert, C.; Richter, I.; Pitz, E.; Planck, B. Astr. Astrophys. Vol. 118, p. 345 (1981).

Leinert, C. In: Properties and interactions of interplanetary dust; Proceedings of the Eighty-fifth Colloquium, Marseille, France, July 9-12, 1984. Dordrecht, D. Reidel Publishing Co., pp. 369-375 (1985).

Leinert, Ch. *et al.* Astronomy and Astrophysics, Vol. 127, p. 1 (1998).

Leisawitz, D.; Hauser, M. G.; Kelsall, T.; Silverberg, R. F.; Odegard, N.; Stemwedel, S.; Weiland, J.; Burdick, S.; Gautier, T. N.; Gillett, F. C.; Murdock, T. L.; Neugebauer, G.; Reach, W. T.; Wheelock, S. New Light on IRAS After DIRBE. American Astronomical Society, 185th AAS Meeting, #57.07; Bulletin of the American Astronomical Society, Vol. 26, p. 1409 (1974).

Levasseur, A. C.; Blamont, J. E. Satellite Observations of Intensity Variations of the Zodiacal Light. Nature, Vol. 246, Issue 5427, pp. 26-28 (1973).

Liou, J.-C.; Dermott, S. F. Dynamical Evolution of Cometary Particles. American Astronomical Society, 24th DPS Meeting, #34.04; Bulletin of the American Astronomical Society, Vol. 24, p. 1005 (1999).

Lissberger, P. H. Sources of error in the modulated wavelength optical thickness monitor for dielectric layers. J. Phys. E: Sci. Instrum., Vol. 2, pp. 875-879 (1969).

Low, F. J. Infrared Cirrus and Zodiacal Dust Bands. Bulletin of the American Astronomical Society, Vol. 16, p. 511 (1984).

Low, F. J.; Young, E.; Beintema, D. A.; Gautier, T. N.; Beichman, C. A.; Aumann, H. H.; Gillett, F. C.; Neugebauer, G.; Boggess, N.; Emerson, J. P. Infrared cirrus - New components of the extended infrared emission. Astrophysical Journal, Part 2 - Letters to the Editor (ISSN 0004-637X), Vol. 278, p. L19-L22 (March 1, 1984).

Madsen, G. J.; Reynolds, R. J.; Ipatov, S. I.; Kutyrev, A. S.; Mather, J. C.; Moseley, S. H. New Observations of the Kinematics of the Zodiacal Dust Cloud. Dust in Planetary Systems. Proceedings of the conference held September 26-28, 2005 in Kaua'i, Hawaii. LPI Contribution No. 1280, p. 111 (2005).

Madsen, G. J.; Reynolds, R. J.; Ipatov, S. I.; Kutyrev, A. S.; Mather, J. C.; Moseley, S. H. New observations of the kinematics of the Zodiacal Dust Cloud. Workshop on Dust in Planetary Systems 2006. Astro-ph/0604229 v1 (2006).

Maris, M.; Burigana, C.; Fogliani, S. Zodiacal Light Emission in the PLANCK mission. Astronomy & Astrophysics, No. 3716 (April 10, 2006).

Meadows, V. S.; Bhattacharya, B.; Reach, W. T.; Grillmair, C.; Noriega-Crespo, A.; Ryan, E. L.; Tyler, S. R.; Rebull, L. M.; Giorgini, J. D.; Elliot, J. L. The Spitzer First Look Survey - Ecliptic Plane Component: Asteroids and Zodiacal Background. Astrophysical Journal Suppl., Vol. 154, p. 469 (2004).

Messenger, S.; Keller, L. P.; Stadermann, F. J.; Walker, R. M.; Zinner, E. Samples of stars older than the sun: Silicate grains in interplanetary dust. Science Express (Feb. 27, 2003).

Minnaert, M.; Mulders, G. F. W.; Houtgast, J. Recensione: Photometric atlas of the solar spectrum, ed: Abetti, G. Memorie della Societa Astronomia Italiana, Vol. 14, p. 72 (1941).

Murdock, T. L.; Price, S. D. Infrared measurements of zodiacal light. Astronomical Journal (ISSN 0004-6256), Vol. 90, pp. 375-386 (Feb. 1985).

Neugebauer, G.; Habing, H. J.; van Duinen, R.; Aumann, H. H.; Baud, B.; Beichman, C. A.; Beintema, D. A.; Boggess, N.; Clegg, P. E.; de Jong, T.; Emerson, J. P.; Gautier, T. N.; Gillett, F. C.; Harris, S.; Hauser, M. G.; Houck, J. R.; Jennings, R. E.; Low, F. J.; Marsden, P. L.; Miley, G.; Olnon, F. M.; Pottasch, S. R.; Raimond, E.; Rowan-Robinson, M.; Soifer, B. T.; Walker, R. G.; Wesselius, P. R.; Young, E. The Infrared Astronomical Satellite (IRAS) mission. Astrophysical Journal, Part 2 - Letters to the Editor (ISSN 0004-637X), Vol. 278, pp. L1-L6 (March 1, 1984).

Nilsson, C. Some doubts about the Earth's dust cloud. Science Vol. 153, pp. 1242-1246 (1966).

Nilsson, C. S.; Southworth, R. B. The Flux of Meteors and Micrometeoroids in the Neighborhood of the Earth. SAO Special Report #263 Bibliographic code: 1967SAOSR.263N (1967).

Over, J. Proc. Kon. Ned. Akad. v. Wetenschapen, Vol. 51B, p. 74 (1958).

Parker, E. N. The Perturbation of Interplanetary Dust Grains by the Solar Wind. Astrophysical Journal, Vol. 139, p. 951 (1964).

Peale, S. J. The Dust Belt of the Earth. Thesis (PhD) Cornell University (1965).

Peterson, A. W. Experimental Detection of Thermal Radiation from Interplanetary Dust. Astrophysical Journal, Vol. 148, p. L37 (1967).

Reach, W. T.; Franz, B. A.; Weiland, J. L.; Hauser, M. G.; Kelsall, T. N.; Wright, E. L.; Rawley, G.; Stemwedel, S. W.; Spiesman, W. J. Observational Confirmation of a Circumsolar Dust Ring by the COBE Satellite. Nature, Vol. 374, No. 6522/APR6, p. 521 (1995).

Reach, W.T. The structured zodiacal light: IRAS, COBE, and ISO observations. ASP conference series, Vol. 124, pp. 33-40 (1997).

Reach, W. T.; Morris, P.; Boulanger, F.; Okumura, K. The mid-infrared spectrum of the zodiacal and exozodiacal light. Icarus, Vol. 164, Issue 2, pp. 384-403 (2003).

Reach, W. T. Spitzer observations of comets and zodiacal light. American Astronomical Society Meeting 206, #28.17; Bulletin of the American Astronomical Society, Vol. 37, p.471 (2005).

Reay, N. K.; Ring, J. Radial Velocity measurements on the Zodiacal Light Spectrum. Nature, Vol. 219, No. 5155, p. 710, (Aug. 17, 1968).

Reay, N. K.; Ring, J. Radial velocity and intensity measurements of the night sky Hβ emission line. Planetary and Space Science, Vol. 17, p. 561 (1969).

Reay, N. K. Zodiacal Dust Particles: Some Comments on Recent Evidence concerning their Motion. Nature, Vol. 224, Issue 5214, pp. 54-55 (1969).

Reynolds, R. J.; Madsen, G. J.; Moseley, S. H. New measurements of the Motion of the Zodiacal Dust. ApJ Vol. 612, pp. 1206-1213 (2004).

Robertson, H. P. Dynamical effects of radiation in the solar system. Monthly Notices of the Royal Astronomical Society, Vol. 97, p. 423 (1937).

Roosen, R. G. The Source of the Gegenschein. Bulletin of the American Astronomical Society, Vol. 2, p. 216 (1970).

Rowan-Robinson, M.; Hughes, J.; Vedi, K.; Walker, D. W. Modelling the IRAS Zodiacal Emission. Monthly Notices of the Royal Astronomical Society, Vol. 246, No. 2, pp. 273-278 (1990).

Rowan-Robinson, M.; Jones, M.; Leech, K.; Vedi, K.; Hughes, J. S. IRAS maps of galactic emission and the zodiacal bands. Royal Astronomical Society, Monthly Notices of the Royal Astronomical Society, Vol. 249, pp. 729-741 (1991).

Sanchez, F. Contribucion al conocimiento del medio interplanetario por fotometria y polarimetria de la Luz Zodiacal. Publicación No. 5, Observatorio Astronomico del Teide, Tenerife, Islas Canarias (1969).

Sanchez Magro, C. PhD Thesis, Luz Zodiacal y Actividad Solar. Tarragona (1974).

Shapiro, I. I.; Lautman, D. A.; Colombo, G. Dynamics of orbiting dust particles. Smithsonian Contributions to Astrophysics, Vol. 11, p. 359 (1967).

Shaw, Frederick G. Comets and their Tails and the Gegenschein Light. Pub: Bailliere, Tindall and Cox (1910).

Singer, S. F. Dust dynamics in the magnetosphere and interplanetary space. Smithsonian Contributions to Astrophysics, Vol. 11, p. 317 (1967).

Singer, S. F. Use of Lunar Data to Determine Space Erosion of Meteorites. Meteoritics, Vol. 4, p. 294 (1969).

Smith, B. and Terrile R. J. A Circumstellar Disk Around Beta Pictoris. Science Vol. 226, No. 4681, pp. 1421–1424 (1984).

Smith, L. L.; Roach, F. E.; Owen, R. W. The absolute photometry of the zodiacal light. Planetary and Space Science, Vol. 13, p. 207 (1965).

Smyth, C. P. "Attempt to apply instrumental measurements to the Zodiacal Light" 20 folios and 2 sheets of drawings. Dated Edinburgh 25 March 1848. Also appendix describing the instrument. Read 25 May 1848. AP.30.18 (location number). Private communication.

Smyth, C. P. Unpublished writings, housed at Edinburgh (c. 1855-60). Private communication.

Smyth, C. P. Teneriffe: An Astronomer's Experiment, or, Specialities of a Residence above the Clouds. Pub: Lovell Reeve, London (1858).

Smyth, C. P. Applying for a grant from the Government Grant of £180 for spectroscopes and £320 for astronomical and geodetical instruments to be used in Egypt, whither he is going to observe the spectrum of the Zodiacal Light and measure the great pyramid. MN.9.269. (location no.) Royal Observatory Edinburgh (25 October 1871). Private communication.

Sparrow J. G ; Ney E. P. Temporal constancy of the Zodiacal Light. Science Vol. 181, pp. 438 440 (1973).

Stark, C. The cosmic infrared background and the importance of zodiacal light modelling. ASTR, Vol. 688, (May 12, 2006).

Sykes, M. V.; Grün, E.; Reach, W. T.; Jenniskens, P. The interplanetary dust complex and comets. Comets II, M. C. Festou, H. U. Keller, and H. A. Weaver (eds.), University of Arizona Press, Tucson, Vol. 745, pp. 677-693 (2004).

Tauber, J, *et al* (eds) PLANCK: The Scientific Programme (The PLANCK Bluebook). ESA-SCI(2005)1, (2005).

Vanysek, V. and Harwit, M. Possibility of a terrestrial component in the Doppler Shifted Zodiacal Light. Nature, Vol. 225, pp. 1231-1232 (March 28, 1970).

Weinberg, J. L. Zodiacal Light at the celestial pole. Planetary and Space Science, Vol. 13, Issue 12, pp. 1311-1312 (1965).

Werner, M. W.; Roellig, T. L.; Low, F. J.; Rieke, G. H.; Rieke, M.; Hoffmann, W. F.; Young, E.; Houck, J. R.; Brandl, B.; Fazio, G. G.; Hora, J. L.; Gehrz, R. D.; Helou, G.; Soifer, B. T.; Stauffer, J.; Keene, J.; Eisenhardt, P.; Gallagher, D.; Gautier, T. N.; Irace, W.; Lawrence, C. R.; Simmons, L.; Van Cleve, J. E.; Jura, M.; Wright, E. L.; Cruikshank, D. P. The Spitzer Space Telescope Mission. The Astrophysical Journal Supplement Series, Vol. 154, Issue 1, pp. 1-9 (2004).

Whipple, F. L. A Comet Model III. The Zodiacal Light. Astrophysical Journal, Vol. 121, p. 750 (1955).

Wolstencroft, R. D. On the location in space of the zodiacal dust particles. Planetary and Space Science, Vol. 15, p. 1081 (1967).